Practical Power Distribution for Industry

Other titles in the series

Practical Data Acquisition for Instrumentation and Control Systems (John Park, Steve Mackay)

Practical Data Communications for Instrumentation and Control (Steve Mackay, Edwin Wright, John Park)

Practical Digital Signal Processing for Engineers and Technicians (Edmund Lai)

Practical Electrical Network Automation and Communication Systems (Cobus Strauss)

Practical Embedded Controllers (John Park)

Practical Fiber Optics (David Bailey, Edwin Wright)

Practical Industrial Data Networks: Design, Installation and Troubleshooting (Steve Mackay, Edwin Wright, John Park, Deon Reynders)

Practical Industrial Safety, Risk Assessment and Shutdown Systems for Instrumentation and Control (Dave Macdonald)

Practical Modern SCADA Protocols: DNP3, 60870.5 and Related Systems (Gordon Clarke, Deon Reynders)

Practical Radio Engineering and Telemetry for Industry (David Bailey)

Practical SCADA for Industry (David Bailey, Edwin Wright)

Practical TCP/IP and Ethernet Networking (Deon Reynders, Edwin Wright)

Practical Variable Speed Drives and Power Electronics (Malcolm Barnes)

Practical Centrifugal Pumps (Paresh Girdhar and Octo Moniz)

Practical Electrical Equipment and Installations in Hazardous Areas (Geoffrey Bottrill and G. Vijayaraghavan)

Practical E-Manufacturing and Supply Chain Management (Gerhard Greef and Ranjan Ghoshal)

Practical Grounding, Bonding, Shielding and Surge Protection (G. Vijayaraghavan, Mark Brown and Malcolm Barnes)

Practical Hazops, Trips and Alarms (David Macdonald)

Practical Industrial Data Communications: Best Practice Techniques (Deon Reynders, Steve Mackay and Edwin Wright)

Practical Machinery Safety (David Macdonald)

Practical Machinery Vibration Analysis and Predictive Maintenance (Cornelius Scheffer and Paresh Girdhar)

Practical Process Control for Engineers and Technicians (Wolfgang Altmann)

Practical Telecommunications and Wireless Communications (Edwin Wright and Deon Reynders)

Practical Troubleshooting Electrical Equipment (Mark Brown, Jawahar Rawtani and Dinesh Patil)

Practical Power Distribution for Industry

Jan de Kock, PR Eng, PH.D, University of Potchefstroom, Potchefstroom, South Africa

Kobus Strauss, CPEng, BCom, B.Eng, SKM Engineers, Perth, Western Australia

Series editor: Steve Mackay

AMSTERDAM • BOSTON • HEIDELBERG • LONDON
NEW YORK • OXFORD • PARIS • SAN DIEGO
SAN FRANCISCO • SINGAPORE • SYDNEY • TOKYO

Newnes is an imprint of Elsevier

ELSEVIER

Newnes

Newnes
An imprint of Elsevier
Linacre House, Jordan Hill, Oxford OX2 8DP
200 Wheeler Road, Burlington, MA 01803

First published 2004

British Library Cataloguing in Publication Data
De Kock, J.
  Practical power distribution for industry. — (Practical professional)
  1. Electric power distribution   2. Industries — Power supply
  I. Title II. Strauss, C.
  621.8'19

Library of Congress Cataloguing in Publication Data
A catalogue record for this book is available from the Library of Congress

ISBN   0 7506 6396 0

For information on all Newnes publications
visit our website at www.newnespress.com

Typeset and edited by Integra Software Services Pvt. Ltd, Pondicherry, India
www.integra-india.com
Printed and bound in The Netherlands

Contents

Preface

This is a practical book on Power Distribution, focusing on medium voltage (1–36 kV) power considerations, switchgear, power cables, transformers, power factor correction, grounding/earthing, lightning protection and network studies. You will gain useful practical technical know-how in these areas, not always covered by university or college programs.

Typical people who will find this book useful include:

- Electrical engineers
- Design engineers
- Project engineers
- Electrical technicians
- Protection technicians
- Maintenance technicians and supervisors.

We would hope that you will gain the following from this book:

- Understand practical power distribution fundamentals
- Determine short-circuit ratings quickly and effectively
- Assess the influence of fault levels on switchgear ratings
- Select the correct type of switchgear for the right application
- Evaluate the advantages of modern state-of-the-art switchgear protection for your applications, including preventative maintenance information
- Recognise the different applications for various cable insulation types
- Know when and how to use single core cables vs three core cables
- Specify correct power cable installation methods
- Correctly utilise and protect power transformers
- Assess and specify correct grounding throughout your electrical network
- Determine the need for Power Factor Correction (PFC) for your environment
- Assess the economic justification for installing PFC equipment
- Correctly specify PFC equipment and be aware of practical consequences
- Confidently use software to solve and predict simple power network problems.

1

Introduction to power distribution

1.1 Introduction

A typical electrical power network is illustrated in Figure 1.1.

An electrical network initiates at the point of generation. Electrical power is generated by converting the potential energy available in certain materials into electrical energy. This is either done by direct conversion of kinetic energy, e.g. wind- or water turbines, or creating steam to drive the turbines, e.g. coal- or nuclear boilers.

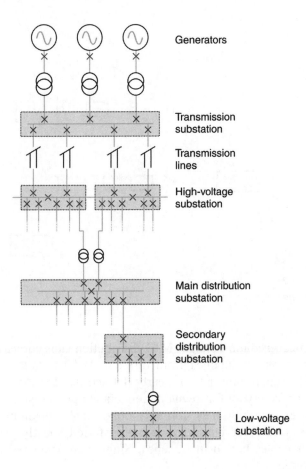

Figure 1.1
Typical electrical power network

The electrical powers generated are either transferred onto a bus to be distributed (small scale), or into a power grid for transmission purposes (larger scale). This is done either directly or through power transformers, depending on the generated voltage and the required voltage of the bus or power grid.

The next step is power transmission, whereby the generated electrical potential energy is transmitted via transmission lines, usually over long distances, to high-voltage (HV) substations. High-voltage substations will usually tap directly into the power grid, with two or more incoming supplies to improve reliability of supply to that substation's distribution network.

Electrical transmission is normally done via high to extra high voltages, in the range of 132–800 kV. Mega volt systems are now being developed and implemented in the USA. The longer the distance, the more economical higher voltages become.

Question 1.1 for course participants: Why would high-voltage transmission be more economical than lower voltages the longer the transmission distance, and be less economical for short distances?

Normally, the transmission voltage will be transformed at the HV substation to a lower voltage for distribution purposes. This is due to the fact that distribution is normally done over shorter distances via underground cables. The insulation properties of three-phase cables limit the voltage that can be utilized, and lower voltages, in the medium-voltage range, are more economical for shorter distances.

Figure 1.2 is a schematical illustration of a typical power grid.

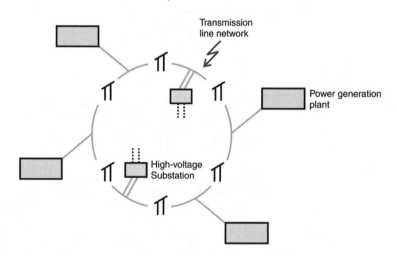

Figure 1.2
Typical power grid

Critical medium-voltage (MV) distribution substations will generally also have two or more incoming supplies from different HV substations. Main distribution substations usually supply power to a clearly defined distribution network, for example, a specific plant or factory, or for town/city reticulation purposes.

Power distribution is normally done on the medium-voltage level, in the range of 6.6–33 kV. Three-phase power is transferred, mostly via overhead lines or 3-core MV power cables buried in trenches. Single-core-insulated cables are also used, although less often.

Low-voltage distribution is also done over short distances in some localized areas.

A power distribution network will therefore typically include the following:

- HV/MV power transformer(s) (secondary side)
- MV substation and switchgear
- MV power cables (including terminations)
- MV/LV power transformer(s) (primary side).

The distribution voltage is then transformed to low voltage (LV), either for lighting and small power applications, or for electrical motors, which is usually fed from a dedicated motor control center (MCC).

This is illustrated in Figure 1.3.

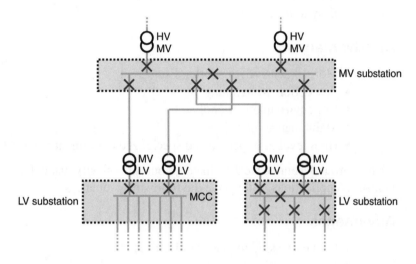

Figure 1.3
Typical power distribution network

Note: Voltage levels are defined internationally, as follows:

- *Low voltage*: up to 1000 V
- *Medium voltage*: above 1000 V up to 36 kV
- *High voltage*: above 36 kV

Supply standards variation between continents by two general standards have emerged as the dominant ones:

- In Europe
  IEC governs supply standards
  The frequency is 50 Hz and LV voltage is 230/400 V
- In North America
  IEEE/ANSI governs supply standards
  The frequency is 60 Hz and the LV voltage is 110/190 V.

Overhead lines are far cheaper than underground cables for long distances, mainly due to the fact that air is used as the insulation medium between phase conductors, and that no excavation work is required. The support masts of overhead lines are quite a significant portion of the costs, that is the reason why aluminum lines are often used instead of copper, as aluminum lines weigh less than copper, and are less expensive. However, copper has a higher current conducting capacity than aluminum per square mm, so once again the most economical line design will depend on many factors.

Overhead lines are by nature prone to lightning strikes, causing a temporary surge on the line, usually causing flashover between phases or phase to ground. The line insulators are normally designed to relay the surge to ground, causing the least disruption and/or damage. This is of short duration, and as soon as it is cleared, normal operation may be resumed. This is why sophisticated auto-reclosers are employed on an increasing number of overhead lines.

Overhead lines have the following properties:

Advantages

- Less expensive for longer distances
- Easy to locate fault.

Disadvantages

- More expensive for shorter distances
- Susceptible to lightning
- Not environment-friendly
- Maintenance intensive
- High level of expertise and specialized equipment needed for installation.

Underground (buried) cable installations are mostly used for power distribution in industrial applications. They have the following properties:

Advantages

- Less expensive for shorter distances
- Not susceptible to lightning
- Environment-friendly
- Not maintenance intensive.

Disadvantages

- Expensive for long distances
- Can be difficult to locate fault.

The focus of this manual will be on MV power distribution, specifically practical considerations regarding MV switchgear, power cables, power factor correction and computer simulation studies.

2

Medium-voltage switchgear

2.1 Standards

Standards in Europe and many other 50 Hz countries are governed by the IEC. The SA bureau of standards SABS has adopted many of the IEC standards e.g. SABS/IEC 87.

In North America, the standards are set by ANSI and the IEEE.

The standards and the way switchgear is tested vary from country to country. It is therefore important to verify what standard is being called for in a specification or is used by the manufacturer. Variations in standards also affect the price of equipment.

2.2 Switch board layout

Switch boards can be designed to have various configurations. This depends on the reliability, fault level and the flexibility required.

Examples:

- Single busbar with only one busbar section (no flexibility and low reliability)
- Single busbar with two sections (improved flexibility and reliability with highest fault level if section maker is closed)
- Single busbar with three sections (excellent flexibility and reliability)
- Double busbar with one or more sections (good flexibility and reliability). The busbar layout may require a break before transfer from one busbar to the next.

2.3 Ratings

Manufactured medium-voltage (MV) switchgear panels are rated according to the following main specifications:

Nominal voltage

This is the designed average voltage of a system, e.g. 11 kV. The actual voltage will typically fluctuate between 10.5 and 11.5 kV (95–105%).

Rated voltage

This is the voltage level at which the equipment will be expected to perform at continuously under normal operating conditions, e.g. 12 kV for a system with a nominal voltage of 11 kV.

The rated voltage will determine the insulation properties of the panel. Each main component, including individual circuit-breakers, busbars, cable terminations, etc., must be rated to this voltage, or higher. ('Main' components refer to those components that form part of the main voltage circuit of the panel, as opposed to the control circuit.) For example, if a 7.2 kV circuit-breaker is installed in a 12 kV panel, it will mean that the whole panel is rated only 7.2 kV.

The normal practice is to rate switchgear panels 10% higher than the required nominal voltage, e.g. 12 kV for an 11 kV system, 36 kV for a 33 kV system, etc.

Power frequency withstand voltage

This is the 50 or 60 Hz voltage that the switchgear can with stand for 1 min. For 12 kV related equipment, the applied test voltage is 28 kV. Switchgear manufacturers use this test to prove the insulation of their equipment after manufacture in routine testing.

Commissioning engineers also use this test to prove the integrity of the equipment before switching on the power. This test is also known as a pressure test.

Impulse voltage

This is the highest peak voltage the equipment will be able to withstand for a very short period of time, as in the case of a voltage peak associated with lightning, switching or other transients, e.g. 95 kV for 12 kV equipment.

BIL standards are set by IEC 60.

The same principle as above applies, i.e. the impulse voltage rating of the panel is equal to the rating of the lowest rated main component.

The impulse level that a panel will experience may be controlled by installing surge suppressers, which will limit the peak voltage to a certain level. This is illustrated in Figure 2.1.

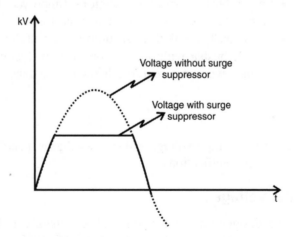

Figure 2.1
Impulse voltage

In limiting the peak voltage, e.g. 45 kV, the surge arrestor will conduct a large current, say 10 kA to earth.

Surge arrestor

Surge arrestors are installed at the transition point between e.g. an overhead line and a transformer, and as close to the transformer terminals as possible.

Surge arrestors are also installed as motor terminals and on the ends of overhead lines. They are unusually connected between phases and earth.

Three types of surge arrestors are used:

1. Rod spark gapped
2. Multiple gapped arrestors
3. Zinc (metal) oxide surge arrestors.

In applying surge arrestors the voltage rating, lighting density and the surge rating should be considered in determining the insulation coordination of the power system.

Full load current

This is the maximum load current that may pass continuously through the switchgear panel. Contrary to the voltage rating, not every main component needs to have the same current rating. Every individual circuit breaker or switch will be rated according to the maximum load current that will pass through it.

The incoming circuit-breaker/disconnector and the main busbars will usually be rated the same. This value is obtained by adding all the individual feeder ratings, multiplied by a load or diversity factor. This factor is smaller than one, and is determined by the types of loads connected to the panel. Individual loads will not run at full capacity simultaneously, hence the load factor.

This calculation can be illustrated as follows:

Feeder No.	Current Rating (A)
1	200
2	100
3	350
4	600
5	150
6	250
Total	1650
Load factor	60%
Required incomer/busbars	990 A

A good estimate of the load factor can be done by looking at the historical values recorded for the relevant feeders, if available.

Due to economy of scale, MV circuit-breakers/disconnectors are manufactured in a few standard sizes, for example 630 A, 1250 A, 1600 A, 2000 A and 2500 A according to IEC standards.

Fault current

The magnitude of fault current that a switchgear panel must withstand is not determined by the load connected to it, but by the properties of the supply to it. Usually, a MV panel will be supplied via a HV/MV transformer(s). This transformer will then determine the magnitude of the fault currents that may flow through the panel.

Electrical faults usually occur due to breakdown of the insulating media between live conductors or between a live conductor and earth. This breakdown may be caused by any one or more of several factors, e.g. mechanical damage, overheating, voltage surges (caused by lightning or switching), ingress of a conducting medium, ionization of air, deterioration of the insulating media due to an unfriendly environment or old age, or misuse of equipment.

Faults are classified into two major groups: symmetrical and unbalanced (asymmetrical). Symmetrical faults involve all three phases and cause severe fault currents and system disturbances. Unbalanced faults include phase-to-phase, phase-to-ground, and phase-to-phase-to-ground faults. They are not as severe as symmetrical faults because not all three phases are involved. The least severe fault condition is a single phase-to-ground fault with the transformer neutral earthed through a resistor or reactor. However, if not cleared quickly, unbalanced faults will usually develop into symmetrical faults. These different types of faults are illustrated in Figure 2.2.

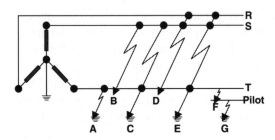

Figure 2.2
Electrical fault types

In distribution systems with Δ/λ (HV/mV) connection and the star point solidly earthed, the single-phase fault current may be higher than the three-phase symmetrical fault current due to the zero sequence reactance of the transformer being lower than the positive sequence reactance.

Switchgear needs to be rated to withstand and break the worst possible fault current, which is a solid three-phase short-circuit close to the switchgear. 'Solid' means that there is no arc resistance. Normally arc resistance will be present, but this value is unpredictable, as it will depend on where exactly the fault occurs, the actual arcing distance, the properties of the insulating medium at that exact instance (which will be changing all the time due to the heating effect of the arc), etc.

Arc resistance will decrease the fault current flowing.

Exact calculations of prospective fault currents can be quite complex, and are usually performed with the aid of computer simulation software (see Chapter 5). However, when a few allowable assumptions are made, approximate fault currents can be determined quite easily and quickly with pen and paper (plus calculator, preferably!). These approximate values will be conservative, giving the worst case, and can therefore be confidently used for the ratings of switchgear panels.

These assumptions are the following:

- Assume the fault occurs very close to the switchgear. This means that the cable impedance between the switchgear and the fault may be ignored.
- Ignore any arc resistance.
- Ignore the cable impedance between the transformer secondary and the switchgear, if the transformer is located in the vicinity of the substation. If not,

the cable impedance may reduce the possible fault current quite substantially, and should be included for economic considerations (a lower-rated switchgear panel, at lower cost, may be installed).

- When adding cable impedance, assume the phase angle between the cable impedance and transformer reactance as zero, hence the values may be added without complex algebra, and values readily available from cable manufacturers' tables may be used.
- Ignore complex algebra when calculating and using transformer internal impedance.

Fault currents are then easily calculated as follows:

Example 1 (ignoring cable impedance)

Ignoring cable impedance means that the fault is actually calculated directly on the transformer's terminals. The magnitude of the fault current is then only limited by the transformer's internal impedance, which is usually indicated by a percentage value on the nameplate of the transformer. This value expresses the impedance value as a percentage of the base value. This is illustrated in the following calculation:

Calculation

Transformer ratings:

Primary voltage:	V_p	$= 132 \text{ kV}$
Secondary voltage:	V_s	$= 11 \text{ kV}$
Capacity:	S	$= 20 \text{ MVA}$
Impedance:	Z	$= 10\%$

Calculation of base value, Z_b:

$$S = \frac{V^2}{Z}$$

$$\Rightarrow Z_b = \frac{V_s^2}{S} \text{ (referred to secondary side of transformer)}$$

$$= \frac{(11 \times 10^3)^2}{(20 \times 10^6)}$$

$$= 6.05 \ \Omega$$

$$Z_{\text{actual}} = Z\% \times Z_b$$

$$= 0.1 \times 6.05 \ \Omega$$

$$= 0.605 \ \Omega$$

$$I_{\text{fault}} = \frac{V_s}{\left(\sqrt{3}.Z_{\text{actual}}\right)}$$

$$= \frac{11 \times 10^3}{1.048 \ \Omega}$$

$$= 10\,496 \text{ A}$$

Manipulation of the above formulas will give a quick reference formula to determine the fault current at the secondary terminals of the transformer:

$$I_{fault} = \frac{S}{\left(\sqrt{3} \times Z\% \times V_s\right)}$$

When more than one transformer is connected in parallel, feeding a switchboard, the fault currents of the transformers are added to give the fault current at the switchboard. For example, if two transformers, with ratings as in the above example, are connected in parallel, the fault current will be 20 992 A.

The smaller the transformer, the smaller the error of this approximation.

Example 2 (with cable impedance)

When the switchgear is situated quite some distance away from the transformer(s), the cable impedance needs to be taken into account to arrive at a realistic value. The fault current will be lower with additional impedance in the circuit, therefore this may be necessary for economic reasons, so as not to overrate new equipment at additional (unnecessary) cost.

The quickest and easiest way to bring cable impedance into the calculation is to simply add the cable impedance, which is usually a Ω/km value in the cable manufacturer's data table, to the transformer reactance (adjusting for line to phase values). This is an approximate value, as complex algebra should be used to add the impedance values, but it is usually accurate enough to determine required switchgear ratings.

Calculation

Take the same transformer ratings as in the above example. Assume the cable distance between the transformer and the switchgear as 6000 m. The cable has the following properties:

Type:	Copper conductors, XLPE insulated
Size:	185 mm^2
Current rating:	410 A
Impedance:	0.1548 Ω/km (see Table 3.28)

Four cables are installed in parallel between the transformer and switchgear to achieve the required current rating.

Transformer reactance
(as per above example): $\quad Z_{Tr} = 0.605$ Ω

Cable impedance:

$$Z_C = 0.1548 \ \Omega \times 6/4$$
$$= 0.2322 \ \Omega$$
$$\Rightarrow Z_{total} = 0.605 \ \Omega + 0.2322 \ \Omega$$
$$= 0.8372 \ \Omega$$
$$I_{fault} = \frac{V_s}{\sqrt{3}.Z_{total}}$$
$$= \frac{11 \times 10^3}{1.450 \ \Omega}$$
$$= 7586 \ A$$

Therefore, if two transformers would feed in parallel to the switchboard in this example, the total fault current would be 15 172 A, and a switchgear panel rated 20 kA would be sufficient; whereas in the previous example a higher rating than 20 kA would be required.

This fault current rating that the switchgear must withstand, is not for an indefinite time, but is normally rated for 1 s or 3 s. The reason for this is that the electrical protection should operate within this time, clearing the fault. Even if the main protection should fail, enough time should be reserved for the back-up protection to function, avoiding severe damage to the switchgear panel. Sufficient time is also allowed for the circuit breaker tripping time.

The fault current rating will then be specified as a current and time rating, e.g. 20 kA, 3 s. This means that the panel will be able to withstand a three-phase through fault current of 20 kA for a continuous period of 3 s.

The fault current rating actually refers to two properties:

1. The amount of electrical energy the panel can absorb, according to the formula, $E = I^2 t$. Exceeding this energy limit will cause thermal damage to the panel, and the conductors will melt.

   Therefore, there is quite a difference between a rating of, for example, 20 kA for 1 s, and 20 kA for 3 s. A panel rated for 20 kA, 1 s will only be able to withstand 11.5 kA for 3 s, according to the following calculation:

$$I_1^2 t_1 = I_2^2 t_2$$
$$\Rightarrow 20^2 \times 1 = I_2^2 \times 3$$
$$\Rightarrow I_2 = \sqrt{(400/3)} \text{ kA}$$
$$= 11.5 \text{ kA}$$

   However, a panel rated for 20 kA, 3 s will not necessarily be able to withstand 34.6 kA for 1 s (according to the same formula), due to the second property involved with fault currents, namely electromechanical stresses.

2. Electromechanical stresses in a switchgear panel arise due to the phenomenon that current flowing in the same direction will cause an electromagnetic field that will assert an attracting force between the current-carrying conductors. In the case of fault currents, this force between the adjacent busbars can be so large that the busbars can be torn loose of their supports, causing extensive damage to the panel.

   This force is proportional to the square of the instantaneous current flowing. Therefore, the mechanical construction of the panel (busbar supports, etc.) will impose an upper limit on its fault current rating, irrespective of the time rating. The peak asymmetrical current rating, including fall DC offset, is used to determine the maximum electromechanical stresses.

It can be seen from the examples that, for MV applications, even long cable distances make only a marginal difference to the fault current rating. This is due to the fact that MV distribution usually involves large cable sizes, with several cables often installed in parallel, with low cable impedance as a result. It is for this reason that the effect of cable impedance is often ignored in MV fault calculations for distribution purposes, except in borderline cases, or where long distances are involved, or in the case of small size cables.

When small MV cables severely reduce the fault current rating, the chances are good that the cable itself will not be able to handle the fault current. This is illustrated in the next chapter on power cables.

Each individual switch as well as the busbars of a switchgear panel must be rated to withstand the prospective fault current. Huge mechanical forces are present in the current-carrying equipment when large fault currents flow, due to electromagnetism. Therefore, the fault current rating do not only refer to the electrical properties of the equipment, but also to their ability to withstand these mechanical forces.

MV switchgear equipment is rated according to the current allowed to pass through it (load or fault current), as well as a 'making' and 'breaking' value. The 'making' current rating is the value of current that the switch may close onto, i.e. the current flowing immediately when the switch is closed. The 'breaking' rating is the value of current already flowing that the switch may interrupt.

Question 2.1 for course participants: Why a 'making' rating, as the current flowing immediately before the switch is closed must be zero (circuit is still open)?

DC offset

The phenomenon called 'DC offset' should be taken into account when rating circuit breakers. This is illustrated in Figure 2.3. This phenomenon occurs due to the presence of inductive reactance in the system.

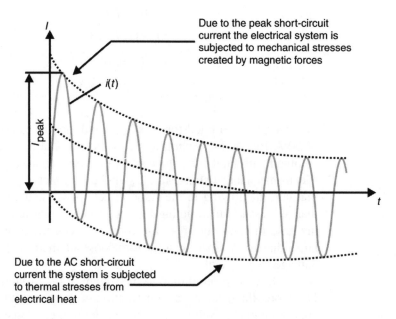

Figure 2.3
Illustration of DC offset

The peak current value will depend on the power factor of the system at the time of the fault. This maximum practical value of this peak current is 2.55 times the RMS fault current, as determined by the system impedance. This relationship is illustrated in Figure 2.4.

Although the peak currents associated with the DC offset is of relative short duration, they must be taken into consideration when rating switchgear panels, as additional mechanical stresses are caused by the electromagnetic forces associated with these high peak currents.

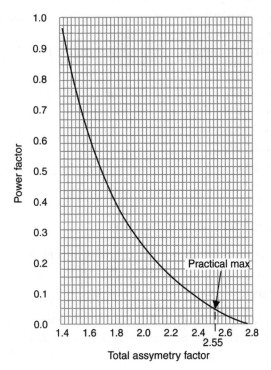

Figure 2.4
Asymmetry factor chart

The *rate of decay* of the DC offset, as illustrated in Figure 2.3, is dependent on the relationship L/R of the system. The higher the inductance, the slower the rate of decay. Normally, the DC offset will decay to zero within the first 3 to 4 cycles. Therefore, at the instant that the circuit breaker opens, these peaks have all but disappeared, and it will not be a cause for concern.

However, when a circuit breaker is installed close to high-inductive sources, like generators and large induction motors, the current waveform may still be substantially asymmetrical at the instant of the circuit breaker opening. This may cause the breaker to interrupt a higher current value than it was rated for, and the DC offset may cause an extended arcing period within the breaker. The resultant thermal energy may be higher than the breaker can withstand, with the result that the breaker may blow up.

Therefore, for these applications, a specialized generator circuit breaker should be installed, which is designed to withstand these factors.

Regulations

The SAA Wiring Rules (Australian Standard AS 3000) stipulates the following for switchgear above 1000 V (Regulation 8.7.2.2):

'The switching device shall be capable of performing at least the following operating functions:

- Make and break full-load current
- Carry the prospective fault current
- Make prospective fault current
- If fitted with protection devices, also break prospective fault current

A manually operated switch need not be capable of breaking the prospective fault current.

A switch which will not make prospective fault current may be approved where satisfactory interlocking is provided.'

Standard fault current ratings for switchgear include:

- 16 kA
- 20 kA
- 25 kA
- 31.5 kA
- 40 kA
- 50 kA
- 60 kA.

2.4   Types of MV switchgear

Switchgear types are classified according to function and method of insulation.

Function

Different terms are used to describe switchgear in different parts of the world. However, the following are most accepted terms recognized internationally.

Switch

A switch is a piece of equipment used to open an electric circuit by a mechanical action, interrupting the flow of current, without causing permanent damage to the equipment when it is used as designed for. A switch may either be a disconnector/isolator or circuit-breaker.

Disconnector/Isolator

The term disconnector or isolator is used to describe a device that is designed to open the electric circuit manually (or by manual command) under normal operating conditions. It may either be an off-load or on-load disconnector/isolator.

'Off-load' means the device is designed to open the circuit without any current flowing. The current rating of an off-load isolator will refer only to the value of current allowed to continuously pass through the device. The making- and breaking current rating will be zero.

'On-load' means the device may open the circuit with normal load current flowing, up to the full load current rating of the device. The continuous- and breaking current ratings will usually be the same, equal to the full load current rating of the device. Modern disconnectors usually have a short-circuit making rating in the region of 25 kA. However, the disconnector will not be able to handle this current for any sustained period of time; hence the current must swiftly be interrupted by a protective device, namely a fuse or a circuit breaker.

Circuit breaker

A circuit breaker (CB) is designed to open an electric circuit under overcurrent or fault conditions. This will usually occur upon a trip signal from a protection relay. A CB will have a continuous full load current rating, a fault current rating for a limited time, as well as a fault making- and breaking rating.

The CB opens its main contacts through a mechanical action. The contacts are kept closed by virtue of mechanical spring pressure. When a trip signal is received, a latch is drawn by a magnetic field set-up by the tripping coil, the spring pressure is released instantly, and the contacts separate.

Due to the mechanical actions involved, a certain fixed amount of time (in ms) elapses from the time the trip command is given until such time as the main contacts actually separate. Additional to this, there is a short period of arcing when the contacts separate, during which time current continues to flow. Therefore, the total tripping (or breaking) time is defined as follows:

$$\text{CB breaking time} = \text{opening time} + \text{arcing time}$$

'Opening Time' represents the time between the instant of application of the tripping signal to the instant of mechanical separation of the main contacts.

'Arcing Time' is the time between the instant of mechanical separation of the main circuit breaker contacts to the instant of arc extinction and hence zero current flow.

This may be illustrated as shown in Figure 2.5.

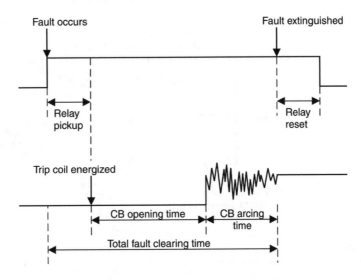

Figure 2.5
Total fault clearing time

Modern circuit breakers have total breaking times between 1 and 5 cycles (40–100 ms).

CB behavior under fault conditions

Before the occurrence of a short circuit, normal load current will be flowing through the CB. This could be regarded as zero when compared to the level of the fault current that would flow. At the instant of the fault, the current will rapidly rise to the maximum value allowed by the network properties. Due to the presence of reactance in the power network, the first few cycles be off-set from the center, oscillating around a fast-decaying axis above the zero line.

This phenomenon is called the DC off-set, or DC transient (see also Section 2.1). The magnitude of the first cycle will depend on network conditions, but may be up to 2.55 times the value of the 'sustained' fault current, i.e. the fault current that will continue to flow until it is interrupted.

The rate of decay of the DC transient will depend on the ratio L/R, i.e. the ratio of system inductance to resistance as seen at the point of the fault.

Under a short circuit or fault condition (Figure 2.6), fault current flows – that has a magnitude which is determined by the network parameters. Upon the breaker tripping, the current is interrupted at the next natural current zero (cyclic point occurring twice a cycle). The network reacts by transient oscillations that give rise to the *transient recovery voltage* (TRV) across the circuit breaker main contacts.

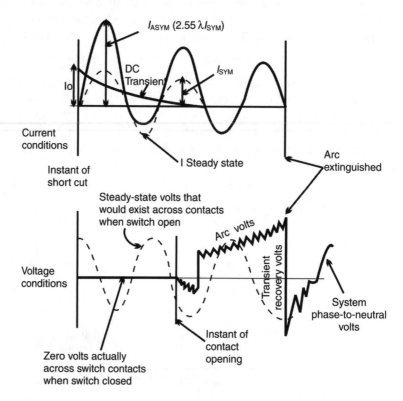

Figure 2.6
Behavior under fault conditions

The insulating medium ensures that the TRV does not cause a flash-over between the contacts and a recurrence of current flow.

The circuit breaker has been designed to withstand and interrupt very high fault currents. Fault current operation places quite high electrical and mechanical stresses on the CB components. For this reason, MV CB life is typically limited to approximately 100 short-circuit operations, depending on the specific type.

Furthermore, circuit breakers are generally limited to around 20 000 load break operations and a maximum of 100 000 mechanical operations.

Contactors

Contactors do not strictly fall under the category 'switchgear', as they are not intended to perform typical switching operations in a power distribution network. Contactors are classified as 'control gear', as their main purpose is to perform control operations, usually starting and stopping of motors, bringing capacitor banks on- and off-line for power factor correction purposes, etc. However, a brief discussion of MV contactors is appropriate here for completeness and comparison purposes.

Vacuum contactors are designed to provide a large number of operations at typical rated loads of up to 400 A per vacuum bottle at voltages of 1.25–13.8 kV. They can typically provide 1–2 million operations at load current. Contactors can be operated in a combination of series and parallel switches to provide the required rating.

It is important to remember that contactors are designed for frequent load control operations, and NOT to interrupt fault currents. Moderate overload currents (up to approximately 120–140%) may be interrupted by a contactor, but certainly not short-circuit currents. Therefore, contactors should not be used for protection purposes, and the contactor itself should be protected against fault currents.

It should be remembered that once a contactor is used to interrupt moderate overload currents, it becomes difficult to prevent the contactor from interrupting fault currents as well, or to ensure that the fault is cleared by other protective devices. Therefore, it is good engineering practice not to use the contactor for protection purposes at all, and to use the contactor in conjunction with other protection devices, like a relay-CB combination or fuses, to clear fault currents.

Due to the fact that the contactor may be utilized for normal load switching and isolation purposes, it is usually an overkill to use a CB in conjunction with a contactor, except in applications where the use of a CB is specifically required. Normally, it is more cost-effective to use fuses to protect the contactor and the relevant part of the network against high overcurrents.

Fuses

Fuses are also not classified as 'switchgear', but rather as protection devices. However, a short discussion of fuses is warranted in the context of electrical distribution equipment.

The fuse is probably the oldest, simplest, cheapest and most-often used type of protection device. The operation of a fuse is very straightforward: The thermal energy of the excessive current causes the fuse-element to melt and the current path is interrupted. Technological developments have served to make fuses more predictable, faster and safer (not to explode).

A common misconception about a fuse, is that it will blow as soon as the current exceeds its rated value (i.e. the value stamped on the cartridge). This is far from the truth. A fuse has a typical inverse time–current characteristic as illustrated in Figure 2.7 and the higher the current, the faster the fuse will blow.

(The curve is typically exponential, due to the equation $e = i^2 t$.)

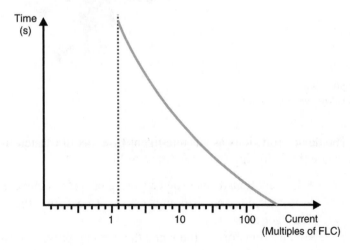

Figure 2.7
Typical fuse characteristic

By nature, fuses can only detect faults associated with excess current. Therefore, a fuse will only blow in earth fault conditions once the current in the faulty phase has increased beyond the overcurrent value. Therefore, fuses do not offer adequate earth fault protection. A fuse has only a single time–current characteristic, and cannot be adjusted. In addition, fuses need to be replaced after every operation. Finally, fuses cannot be given an external command to trip.

Fuses are inexpensive. Therefore, they are suitable to use solely on less critical circuits, in conjunction with contactors or as back-up protection should the main protection fail. They offer very reliable current-limiting features by nature.

Another advantage of fuses is the fact that they can operate totally independently, i.e. they do not need a relay with instrument transformers to tell them when to blow. This makes them especially suitable in applications like remote Ring Main Units, etc.

I_S-limiter

A very 'special' type of fuse is the I_S-limiter, originally developed by the company ABB. The device consists of two main current-conducting parts, namely the main conductor and the fuse, as illustrated in Figure 2.8.

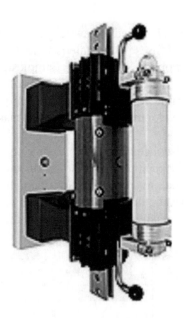

Figure 2.8
Construction of I_S-limiter

The device functions as an 'intelligent fuse', as illustrated in Figure 2.9. The functional parts are the following (with reference to Figure 2.9):

- Current transformer (detects the short-circuit current)
- Measuring and tripping device (measures the current and provides the triggering energy)
- Pulse transformer (converts the tripping pulse to busbar potential)
- Insert holder with insert (conducts the operating current and limits the short-circuit current).

Figure 2.9
Functional diagram of I_S -limiter

The I_S-limiter is intended to interrupt very high short-circuit currents very quickly, in order to protect the system against these high currents. Currents of up to 210 kA (11 kV) can be interrupted in 1 ms. This means that the fault current is interrupted very early in the first cycle, as illustrated in Figure 2.10.

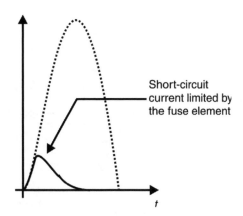

Figure 2.10
Fault current cycle

When a fault current is detected, the main conductor is opened very swiftly. The current then flows through the fuse, which interrupts the fault current. The overvoltage occurring due to the interruption of current is relatively low due to the fact that the magnitude of current on the instant of interruption is still relatively low. The main conductor and parallel fuse have to be replaced after each operation.

The I_S-limiter is only intended to interrupt high fault currents, leaving the lower fault currents to be interrupted by the circuit breakers in the system. This is achieved by the measuring device detecting the instantaneous current level and the rate of current rise. The rate of current rise (di/dt) is high with high fault currents, and lower with lower fault currents, as illustrated in Figure 2.11. The I_S-limiter only trips when both set response values are reached.

A practical use of the I_S-limiter is illustrated in Figure 2.12, where the combined fault current supplied by two transformers in parallel would be too high for the switchgear panel to withstand.

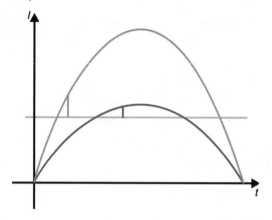

Figure 2.11
Rate of current rise

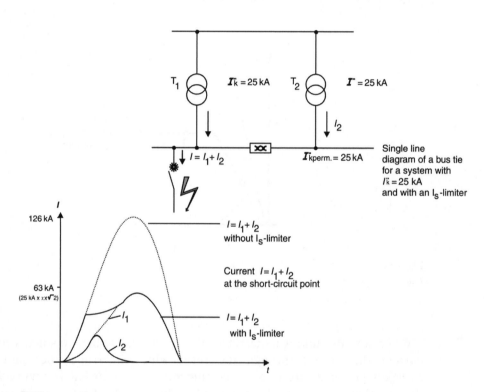

Figure 2.12
Practical use of I_S-limiter

2.5   Insulation methods

Oil circuit breakers

In this design, the main contacts of the circuit breaker are immersed in oil.

Oil in CB has two functions i.e. cooling the arc and providing insulation of the live parts.

The energy inherent in the electrical arc has the effect of decomposing the switch oil into four component parts:

1. 70% hydrogen
2. 22% acetylene
3. 5% methane
4. 3% ethylene.

Due to this decomposition into gas components, the arc effectively lives in a bubble of gases, this being surrounded by the oil.

The formation of gas and carbon during arcing reduces the insulation properties of the oils. This requires the oil to be charged on a regular basis.

Oil has the following advantages

- Ability of cool oil to flow into the space after current zero and therefore extinguishing the filament arc that is still in existence
- A cooling surface is presented by oil to the contact surfaces
- Absorption of arc-energy by decomposition of oil from liquid to gas
- Oil acts as an additional insulator lending to more compact designs of switchgear.

Disadvantages

- Inflammability, especially if there is any O2 near the generated gasses
- Additional maintenance is required to ensure that the oil is kept in a suitable and adequate operational state
- Additional fire and environmental hazard with possible leakage of oil out of switchgear.

Oil circuit breakers are not manufactured on a large scale anymore by the world's leading switchgear manufacturers. This is mainly due to the major disadvantages as mentioned above, namely fire and environmental hazard, and high maintenance requirements.

Air break switchgear

Air break switchgear can be divided into two major divisions, namely air circuit breakers (ACBs), mainly utilized for low-voltage applications, and high-voltage air disconnectors and breakers utilized in high-voltage switchyards. The latter falls outside the scope of this manual. The ACB is discussed briefly.

Air circuit breaker (ACB)

The interrupting contacts are situated in free air, and as such use this medium for arc suppression and cooling. This is illustrated in Figure 2.13. The ACB normally uses an electronic tripping device, as an integral part of the equipment, to provide the tripping signal.

The resultant arc is chopped into a number of smaller arcs by the arc-chute as it rises due to the generated heat and associated magnetic forces. In some cases air puffers are used to aid the upward movement of the arc. In other cases air under high pressure is used to 'blast' the arc away. This type of breaker is relatively large, due to the distance needed between phases with only air as insulating medium. New air breakers are very seldom used for voltages exceeding 3.3 kV nowadays.

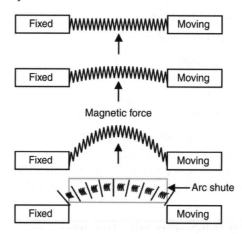

Figure 2.13
Illustration of ACB contacts

The following are the advantages and disadvantages of air breakers.

Advantages

- Economical choice for voltages up to 1000 V
- Simple construction
- No fire or health hazard
- 'Soft' break of current, due to the fact that the ionization of air will support current flow during the early stages of contact separation
- High short-circuit withstand and breaking capability.

Disadvantages

- Not economical for voltages exceeding 3.3 kV (too big)
- Requires regular maintenance in mV application's.

Sulphur-hexaflouride (SF$_6$)

Sulphur-hexaflouride (SF$_6$) is an inert insulating gas that is becoming more and more popular in modern switchgear designs both as an insulating as well as an arc-quenching medium.

The pressure of the SF$_6$ gas is generally maintained at a pressure between 2 and 10 bar. So good sealing of the gas chambers is vitally important. Leaks will cause loss of insulating medium and it has to be remembered that contact clearances are not designed for operation in air. Certain types of switchgear are rated to be able to break normal load current in the absence of SF$_6$, i.e. with only air as insulating medium, *but not fault current*. Opening of the contacts in air under short-circuit conditions will have catastrophic consequences.

The SF$_6$ switchgear can have the facility to continuously monitor the gas pressure in the contact chambers. Modern relays have the capability to automatically raise an alarm the instant this pressure falls below a specified limit, trip or lock-out the CB in its current position when a second predefined pressure limit has been reached, and/or continuously send through gas pressure information to a SCADA control station (see Chapter 6).

Even though the risk of gas leaks is very low with modern switchgear, the monitoring of SF$_6$ gas pressure is strongly recommended, either automatically or as part of a manual maintenance procedure.

Practical tip

A common concern regarding the use of SF_6 gas pressure monitors are that the monitor itself posses the highest risk of causing a leakage. This very seldom occurs when switchgear is manufactured under tight quality control measures, as it should be. However, it can be seen as an additional point of risk. But even if a gas leakage develops at the pressure monitor, at least an alarm will be raised, and the responsible personnel will know about it, and corrective measures can be taken in advance.

With no monitor installed, no one will know about a gas leakage until it is too late, with possible disastrous consequences. Therefore, having pressure monitors on the switchgear can be regarded as the lesser of two evils by far.

SF_6 gas

SF_6 gas is an inert gas, colorless and odorless. It is not poisonous or flammable. It is heavier than air, hence in case of a leakage, SF_6 gas will collect in the lowest regions of the substation, which are normally the cable trenches, and replace all the air. Therefore, it may pose a health risk to human beings in that it may cause suffocation to unaware workers.

Question 2.2 for course participants: How would you prevent a possible build-up of SF_6 *gas when using* SF_6 *type switchgear?*

GIS switchgear

Gas-Insulated switchgear (GIS) also uses SF_6 gas to insulate not only the switch contacts, but most of the switchgear panel, including the busbars. The GIS has been used at the higher-voltage levels, namely 66 kV and above, but is also becoming a viable option for medium-voltage applications in special circumstances. The switchgear manufacturers ABB, for example, has used this technology very successfully in designing efficient, compact and safe Ring Main Units.

The main advantage of GIS is the reduced clearance distances required due to the superior insulating properties of SF_6 gas, resulting in compact, space-saving switchgear. Therefore, the higher the voltage level, the more the space-saving benefits obtained.

The following are the advantages and disadvantages of SF_6.

Advantages

- 'Softer' switching than, for example, vacuum breakers, meaning that less pronounced voltage peaks are caused when interrupting large currents
- Suitable for voltages up to 66 kV for metal-clad switchgear and 800 kV for GIS
- SF6 gas pressure can be easily monitored
- Insulating medium present no fire/explosion hazard.

Disadvantages

- Depositories formed during switching limit the number of achievable operations before maintenance/refurbishment is required
- More expensive than other types of switchgear
- Decomposed SF6 products are toxic and should be handled with care.

Vacuum breakers

Vacuum switchgear applies the principle that it is virtually impossible for electrical current to flow in a vacuum. Therefore, the main contacts separate inside a vacuum bottle.

The typical contact material comprises a tungsten matrix impregnated with a copper and antimony alloy to provide a low melting point material to ensure continuation of the arc until nearly current zero.

The early designs of vacuum breakers displayed the phenomenon of current chopping i.e. switching off the current at a point on the cycle other than current zero. This sudden instantaneous collapse of the current generated extremely high-voltage spikes and surges into the system, causing failure of equipment (Remember the formula $V = L \, \mathrm{d}i/\mathrm{d}t$?).

Another phenomenon was prestrike at switch on. Due to their superior rate of dielectric recovery, a characteristic of all vacuum switches was the production of a train of pulses during the closing operation. Although of modest magnitude, the high rate of rise of voltage in prestrike transients can, under certain conditions, produce high insulation stresses in motor line end coils.

Subsequent developments attempted to alleviate these shortcomings by the use of 'softer' contact materials, in order to maintain metal vapor in the arc plasma so that current continues to flow until point zero. Unfortunately, this led to many instances of contacts welding on closing.

Re-strike transients produced under conditions of stalled motor switch-off were also a problem in vacuum contactors. When switching off a stalled induction motor, or one rotating at only a fraction of synchronous speed, there is little or no machine back emf, and a high voltage appears across the gap of the contactor immediately after extinction. If at this point in time the gap is very small, there is the change that the gap will break down and initiate a re-strike transient, puncturing the motor's insulation.

Modern designs have all but overcome these problems. In vacuum contactors higher operating speeds coupled with switch contact material have chosen to ensure high gap breakdown strength, produce significantly shorter trains of pulses.

In vacuum circuit breakers, operating speeds are also much higher which, together with contact materials that ensure high dielectric strength at a small gap, have ensured that prestrike transients have ceased to become a significant phenomenon.

However, by nature of their design, vacuum switchgear do cause higher-voltage peaks when breaking high currents, compared to other types of switchgear. With the use of special surge arrestors, e.g. ZORCs* and wave-slowing capacitors, the transient overvoltages can be prevented from damaging the motor's insulation. (*a registered trademark of Strike Technologies).

The following are the advantages and disadvantages of vacuum switchgear.

Advantages

- Due to virtually no depositories being formed during the breaking operation, vacuum switchgear support a large number of operations
- More compact than the other types
- Vacuum bottles are relatively inexpensive and easy to replace
- Very little maintenance needed.

Disadvantages

- Tendency to cause voltage 'spikes' when interrupting large currents
- No practical method available at this stage to monitor the vacuum inside the bottle
- Generally limited to voltages of up to 36 kV.

Figures 2.14 and 2.15 illustrate the typical construction of a vacuum circuit breaker.

Comparison of insulating methods for CBs

Property	Air	Oil	SF$_6$	Vacuum
Number of operations	Medium	Low	Medium	High
'Soft' break ability	Good	Good	Good	Fair
Monitoring of medium	N/A	Manual test	Automatic	Not possible
Fire hazard risk	None	High	None	None
Health hazard risk	None	Low	Low	None
Economical voltage range	Up to 1 kV	3.3–400 kV	3.3–800 kV	3.3–36 kV

Figure 2.14
Typical vacuum circuit breaker

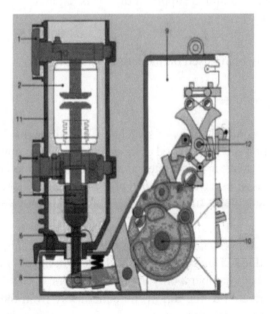

1 Upper connection
2 Vacuum interrupter
3 Lower connection
4 Roller contact (swivel contact for 630 A)
5 Contact pressure spring
6 Insulated coupling rod
7 Opening spring
8 Shift lever
9 Mechanism housing with spring operating mechanism
10 Drive shaft
11 Pole tube
12 Release mechanism

Figure 2.15
Internal construction of a vacuum circuit breaker

2.6   Types of closing mechanisms

Hand operated

A cheap version but is losing popularity as the operating speed depends entirely on the operator's dexterity. Additionally it is only used at ratings of 150 MVA and less, and up to a maximum of 11 kV.

Hand operated and spring assisted

Hand actuator compresses a cam-spring arrangement over top dead center (>90°). The spring is under compression and, once past T.D.C, completes the cycle to 180° and closes the breaker.

Quick make

A variance on the device described in version (2) above. A spring is compressed and latched by a hand-operated system. The spring is then released by operation of the latch to operate mechanism.

Motor wound spring

Versions (2) and (3) above can have a motor replacing the manual function as described. Motor-driven systems have gearboxes, and are used for larger systems (high MVA ratings).

Solenoid

Versions (2) and (3) above can have a solenoid replacing the manual function as described; this is more common on the smaller ranges of switchgear.

Pneumatic

Used at 69 kV and above. Convenient, but a supply of dry air is required.

Hydraulic

Hydraulic pressure is increasingly being used on modern circuit breakers.

2.7   Internal arc proofing

Internal arc proofing refers to the certified ability of switchgear to withstand an internal short circuit, typically a busbar fault, for a specified period of time without causing injury to persons standing directly in front of the switchgear panel. Internal arc proofing need to be certified by an authorized testing authority, and will state a current and time rating, e.g. 25 kA, 1 s.

The following will typically be features of an internal arc proof panel for 200 ms:

- Mild steel of at least 3 mm thickness for panel construction
- Secured, bolted front doors. The door to the CB enclosure cannot be opened while the CB is racked in. CB needs to be racked out with the door closed
- No easy access to busbar chambers, only via bolted covers
- Arc energy will be directed to the top of the panel away from the front doors and 'blown out' at the top via false covers.

It is theoretically safe to stand in front of a panel that is being faulted.

Because the internal arc proofing is short time rated, bus zone protection is required to reduce the fault clearing time. Examples of bus zone protection include:

- High- and low-impedance differential protection schemes (operates in less than 10 ms)
- Busbar blocking schemes (operates in 100 ms or more)
- Arc detection schemes (operates in less than 20 ms).

2.8   Modern protection relays used with switchgear

Although this is not a text on electrical protection, a short discussion on modern protection relays are justified, as modern relays are changing the way electrical engineers think, not only about protection, but regarding a much wider scope, including switchgear applications and maintenance issues.

The most advanced protection relays available today are termed IEDs (intelligent electronic devices), as they offer much more than only the protection functions of the traditional relay.

Functions of an IED

The functions of a typical protection IED can be classified into five main areas, namely protection, control, monitoring, metering and communications. Some IEDs may be more advanced than others, and some may emphasize certain functional aspects over others, but these main functionalities should be incorporated to a greater or lesser degree.

Protection

The protection functions of the IED evolved from the basic overcurrent and earth fault protection functions of the feeder protection relay (hence certain manufacturers named their IEDs 'feeder terminals'). This is due to the fact that a feeder protection relay is used on almost all cubicles of a typical distribution switchboard, and the fact that more demanding protection functions are not required enable the relay's microprocessor to be used for control functions. The IED is also meant to be as versatile as possible, and is not intended to be a specialized protection relay, for example generator protection. This also makes the IED affordable.

The following is a guideline of protection-related functions that may be expected from the most advanced IEDs (the list is not all-inclusive, and some IEDs may not have all the functions). The protection functions are typically provided in discrete function blocks, which are activated and programed independently.

- Non-directional three-phase overcurrent [low-set, high-set and instantaneous function blocks, with low-set selectable as long time-, normal-, very-, or extremely inverse, or definite time]
- Non-directional earth fault protection [low-set, high-set and instantaneous function blocks]
- Directional three-phase overcurrent [low-set, high-set and instantaneous function blocks, with low-set selectable as long time-, normal-, very-, or extremely inverse, or definite time]
- Directional earth fault protection [low-set, high-set and instantaneous function blocks]

- Phase discontinuity protection
- Three-phase overvoltage protection
- Residual overvoltage protection
- Three-phase undervoltage protection
- Three-phase transformer inrush/motor start-up current detector
- Auto-reclosure function
- Underfrequency protection
- Overfrequency protection
- Synchro-check function
- Three-phase thermal overload protection.

Control

Control functions include local and remote control, and are fully programmable.

- Local and remote control of up to six switching objects (open/close commands for circuit-breakers, isolators, etc.)
- Control sequencing
- Bay level interlocking of the controlled devices
- Status information
- Information of alarm channels
- HMI panel on device.

Monitoring

Monitoring includes the following functions:

- Circuit-breaker condition monitoring, including operation time counter, electric wear, breaker travel time, scheduled maintenance
- Trip circuit supervision
- Internal self-supervision
- Gas density monitoring (for SF6 switchgear)
- Event recording
- Other monitoring functions, like auxiliary power and relay temperature.

Metering

Metering functions include:

- Three-phase currents
- Neutral current
- Three-phase voltages
- Residual voltage
- Frequency
- Active power
- Reactive power
- Power factor
- Energy
- Harmonics
- Transient disturbance recorder (up to 16 analog channels)
- Up to twelve analog channels.

Communications

Communication capability of an IED is one of the most important aspects of the device today and is the one aspect that clearly separates the different manufacturers' devices from one another regarding their level of functionality.

IEDs are normally able to communicate directly to a SCADA system, i.e. upper level communications. Different manufacturers use different communication protocols, and this will determine certain critical aspects, e.g. response times, possibility of direct relay-to-relay communications, etc.

An IED will, in addition to upper level communications, also have a serial port or optical interface to communicate directly to substation PC or laptop for configuration and data downloading purposes, should the SCADA link not be available or desirable in that instance.

Comparison of electromechanical relays and digital relays

No.	Feature	Electromechanical	Digital
1	Reliability	Moderate	High
2	Stability	High	High
3	Sensitivity/accuracy	Low	High
4	Speed of operation	Moderate	High
5	Discrimination capability	Moderate	High
6	Multi-function	No	Yes
7	Versatile (can be used for different applications)		Yes
8	Flexible (multiple curves, selectable setting groups)	No	Yes
9	Maintenance intensive	High	Low
10	Self-diagnostics	No	Yes
11	Trip circuit supervision	No	Yes
12	Condition monitoring	No	Yes
13	Data communications	No	Yes
14	Control functions	No	Yes
15	Metering	No	Yes
16	Disturbance recordings	No	Yes
17	Remote operation	No	Yes
18	CT Burden	High	Very low
19	Cost	Low	See Note 1

Note 1

The cost will depend on the application. If a multi-function relay, like an intelligent feeder terminal, is used for only overcurrent protection, the cost for that function will be high. However, if a digital relay is fully utilized, for example a generator protection relay, the cost per function will be low.

Note 2

Not all digital relays will comply with the features as in the table, depending on the type of relay. However, the table is an overview of what is possible with digital relays in general, compared to what could be achieved with electromechanical relays.

3

Power cables

3.1 Introduction

The following types of power cable are mainly used for distribution purposes in the medium-voltage (MV) range and will be the focus of discussion in this chapter:

- Paper-insulated lead covered (PILC) cables
- Cross-linked polyethylene (XLPE) cables.

PVC cables are used mainly in the low-voltage (LV) range, and will be briefly discussed, as well as elastomeric trailing cables, which are used for specialized applications, for example the mining industry.

Overhead lines are mainly used for transmission purposes and will only be briefly mentioned.

IDC acknowledges the contribution of Aberdare cables and in particular the efforts of Dick Hardy.

3.2 Paper-insulated lead-covered (PILC) cables

The PILC cables are manufactured by using layers of paper impregnated with a compound mineral oil as insulating medium, both as individual core and overall insulation. A lead sheath is constructed as an outer core layer to mainly provide a seal for the compound in the paper layers, and also for excellent corrosion protective properties as well as to provide additional mechanical protection.

A steel tape layer (often a double layer) or steel wires are used for the main mechanical protection and it may also be used as a return path for earth currents. The outer sheath may be a PVC layer or other type of insulating and waterproof material.

DSTA = Double steel tape armored

SWA = Steel wire armored

Earlier PILC cables displayed the phenomenon of compound migration, meaning the compound tended to drain toward the lower part of the cable when installed vertically or against slopes. Non-draining compound are used in modern manufacturing techniques, eliminating this phenomenon.

PILC cables are generally used for 66 kV up to 33 kV applications. The designation 6.35/11 kV, for example, means that the cable has an insulation level of 6.35 kV between each core and earth, and 11 kV between phases. Cables rated 6.35/11 kV may be used for earthed systems (transformer neutral earthed), whereas 11/11 kV rated cables should be used for unearthed systems. The same applies for other voltage levels.

Detail specifications of PILC cables are given in Section 3.8 and Table 3.27.

3.3   Cross-linked polyethylene (XLPE) cables

XLPE, PEX and PVC are used as conductor insulating materials in these cables. XLPE is a semiconductor, and provide partial insulation as well as electrical stress relieving. The conductors, with their XLPE layers, are embedded in PVC to provide total insulation. Steel wires are used for mechanical strength, and may also be used to provide the return path (or part thereof) for earth fault currents. The outer sheath is normally a PVC sheath (Figure 3.6), to provide insulation and waterproofing.

XLPE cables are used from low voltage (600/1000 V) to 132 kV applications. Pure PVC cables, in which PVC replaces the XLPE as conductor layer, are used up to a maximum of 6.6 kV.

Detail specifications of XLPE and PVC cables are given in Section 3.8 and Table 3.28.

3.4   Elastomeric cables

These types of cables use a rubber-type elastomeric material for insulation to provide mainly greater flexibility as well as high corrosion resistance against harsh environmental conditions and ultraviolet radiation. The outer sheath is usually reinforced with a mesh braiding.

Elastomeric cables are mainly used as trailing cables in mining operations and other specialized applications.

3.5   Aerial bundled conductors (ABC)

This type of cable is used for LV distribution of power in residential areas. The three insulated conductors are rapped around a steel rope and strung between overhead live poles. This reduces the risk of open conductors touching, and trees causing faults.

3.6   Overhead aluminum conductors

Aluminum conductors have gained wide acceptance all over the world for use in overhead transmission and distribution lines. Generally a steel core is used with the aluminum to give the conductor mechanical strength, termed aluminum conductor steel reinforced (ACSR).

Aluminum is preferred as conducting medium in overhead lines above copper, mainly based on economic considerations. Aluminum is considerably lighter than copper, even when allowing for less current density of aluminum compared to copper. This has a considerable economic advantage when designing overhead lines, as span lengths can be greater and support masts lighter.

Aluminum also has superior corrosion-resistant properties, which is a great benefit for overhead lines, as no insulating material is present to protect the conductors against corrosion, as is the case in cables.

The design and application of overhead lines are very specialized, and they are nowadays used mainly for transmission purposes. Therefore, a detail discussion of overhead lines falls outside the scope of this manual.

3.7   Cable selection

When selecting a cable for your specific application, a number of variables require attention. These are:

- Application
- Size and type of load to be supplied
- Permissible voltage drop
- Prospective fault current
- Circuit protection
- Installation conditions.

Application

Low-voltage distribution

For LV distribution purposes, the choice is basically between XLPE and PVC-insulated cables. The XLPE cables have higher current ratings than PVC cables for the same conductor size, as can be seen from Tables 3.25 and 3.26. Normally XLPE cables tend to be slightly more expensive than PVC cables.

The choice between these two types for LV applications will normally be determined by economic considerations (the relative prices at that stage) and availability. Bear in mind that a slightly smaller XLPE cable can be chosen for the same current requirement, which have other spin-offs, for example space-saving on cable racks or in trenches, slightly reduced labor costs for installation, etc.

Medium-voltage distribution

For MV applications, the choice is more involved. First of all, the choice lies between overhead lines and underground cables. Nowadays, the tendency is to move toward underground cables for distribution purposes, the motivation mainly being the following:

- Distribution is mainly done in industrial and/or populated areas, where overhead lines pose an environmental hazard.
- Overhead lines are very maintenance intensive, due to required protection against corrosion, pollution and fires.
- Overhead lines are very susceptible to lightning strikes.

Due to the above considerations, overhead lines are mainly considered for applications over relatively large distances, where they are very economic compared to underground cables.

The next choice is then between aluminum or copper conductors. Aluminum conductors are larger than copper conductors for the same current-carrying capacity, which may add to installation costs. The choice will mainly be an economic one, influenced by availability and the relative prices of the two metals at that stage. Aluminum conductors may be considered in high corrosive areas.

The third choice is the question of cable insulation type. For normal distribution purposes, the choice lies between PILC- or XLPE-insulated cables. The choice will be influenced by taking the following factors into account:

- PILC cables tend to be more expensive than XLPE cables for the same conductor size.
- XLPE cables have a higher current rating for the same conductor size.

- XLPE cables have a larger overall diameter for the same conductor size.
- PILC cables provide better corrosion resistance than XLPE cables.
- PILC cables have a higher average life span, 40 years compared to 25 years of XLPE cables.
- XLPE cables are better suited to be moved frequently after installation, making them better suited for use in an continuously changing environment, for example mining applications.

Single-core cables may also offer an economical choice, especially in high-current applications, where the installation of three single-core cables (one per phase) might be preferred to the option of installing a number of 3-core cables in parallel.

The outer 'PVC' sheath is available in finishes, e.g:

- No stripe – plain PVC
- Red stripe – fire retardant
- Blue stripe – low halogen
- White stripe – zero halogen.

All 'PVC' covering are also UV resistant.

Once a choice is made, companies tend to stick to a specific type of cable, in order to prevent confusion, eliminate the need for re-training of personnel, and reduce stock levels.

Load to be supplied

In order to select the appropriate cable, it is necessary to know the voltage and the load current, as the first step in the selection process.

The following formulae apply:

$$I_{FL} = \frac{kW \times 1000}{\sqrt{3} \times V \times \cos \phi} \text{ A if we know kW, voltage, and power factor}$$

$$= \frac{kVA \times 1000}{\sqrt{3} \times V} \text{ A if we know the kVA rating and voltage}$$

Use this value of current to determine the cable size by reference to the relevant manufacturer's tables for copper or aluminum conductors.

A slightly larger conductor size may be chosen for safety aspects, and to provide for the higher than usual current, which may be experienced during starting of electric motors.

Example of cable selection for low voltage

(*Note*: The tables used in this manual are for illustrative purposes only. They are based on data supplied by the cable manufacturer Aberdare Power Cables. The user should use the corresponding tables supplied by the specific manufacturer of the cable he is using, as each manufacturer's specifications will vary.)

Suppose it is required to supply a three phase, 400 V, 100 kW motor, over a distance of 50 m, the motor load is known to have a power factor of 0.9 lagging. The full load line current, I_{FL} can be calculated as follows:

$$I_{FL} = \frac{100 \times 1000}{\sqrt{3} \times 400 \times 0.9}$$

$$= 161 \text{ A}$$

Full load current ratings can usually be obtained easily from the equipment nameplate.

We now refer to Table 3.25 (example table for LV cables) and note that the smallest copper conductor, PVC-insulated cable, that can supply a current of 161 A in air, is a 50 mm^2 rated area cable. This cable can carry 167 A continuously if installed under standard conditions.

Permissible voltage drop

Calculate the highest current drawn by the load, by multiplying the current as calculated in Section 3.7 by an appropriate factor. If a Star/Delta motor starter is used on a motor, this factor is 3. If the motor is started direct on line, then use a factor of 6. A factor of 2 may be used for most electronic soft-starters. Where the load is resistive heating, lighting or a transformer, it is not necessary to increase the current as calculated in Section 3.7.

Calculate the volt drop that will be experienced at the load terminals. The maximum volt drop allowed by the Australian Standard for Electrical Installations (Standard AS 3000; known as the SAA Wiring Rules) is 5%.

The volt drop may be calculated in two different ways:

1. Multiplying the current by the impedance of the length of cable. Calculate the percentage volt drop by reference to the phase-to-earth voltage.
2. Multiply the current by the length of cable, and then multiply the result by the volt drop per amp per meter figure as given in Table 3.25.

Previous example using method (a)

$$\text{Starting current} = 3 \times \text{running current}$$
$$= 3 \times 161\,\text{A}$$
$$= 483\,\text{A}$$

$$\text{Impedance of 50 m of 50 mm}^2 \text{ cable (Table 3.25)} = \frac{0.4718}{1000} \times 50$$
$$= 0.02359\,\Omega$$

$$\text{Volt drop} = 483 \times 0.02359$$
$$= 11.394\,\text{V}$$

$$\text{Percentage volt drop} = \frac{11.394}{230} \times \frac{100}{1}$$
$$= 4.95\%\ \text{(acceptable)}$$

Previous example using method (b)

$$\text{Starting current} = 3 \times 161\,\text{A}$$
$$= 483\,\text{A}$$

$$\text{Volt drop per amp per meter} = 0.817\,\text{mV/A/m (Table 3.25)}$$

$$\text{Volt drop} = 0.817 \times 10^{-3} \times 483 \times 50$$
$$= 19.73\,\text{V}$$

$$\text{Percentage volt drop} = \frac{19.73}{400} \times \frac{100}{1}$$
$$= 4.93\%\ \text{(acceptable)}$$

Note: It often happens on long runs of electric cable that a larger conductor than that calculated in 3.7 is required for volt drop reasons.

It must be noted that the calculation as per the above example is an approximation. However, it is accurate enough for cable selection purposes. (In fact, it is accurate enough for most practical purposes.)

Question 3.1 for course participants: Why is the method as illustrated above only an approximation? What is the correct method to calculate the exact volt drop?

Example of cable selection for MV (11 kV)

We wish to supply a 2 MVA 11 kV transformer from a utility supply which is 3 km away. Underground paper-insulated, copper conductor cable is to be used.

The depth of burial of the cable is 1.25 m. Ground thermal resistance is 2 Km/W. The ground temperature is 25 °C and there are no other cables in the trench.

Short-circuit level may be assumed to be 250 MVA, and the earth fault level 100 MVA, and it may be assumed that a fault will be cleared in half a second.

Using the previous formulae:

$$I_{FL} = \frac{2\,000\,000}{\sqrt{3} \times 11\,000}$$

$$= 105 \text{ A before derating for non-standard conditions}$$

De-rating factor for depth of burial at 1.25 m is 0.96 (see Section 3.8).
De-rating factor for soil thermal resistivity at 2 Km/W is 0.84.
De-rating factor for ground temperature of 25 °C is 1.00.

$$\text{Total derating} = 0.96 \times 0.84 \times 1.00$$
$$= 0.806$$

$$I_{standard} = 105/0.806$$
$$= 130 \text{ A}$$

Table 3.27 shows that a 35 mm^2 copper conductor cable would be capable of carrying this load (130 A).

Checking for volt drop

$$V_{drop} = \frac{\sqrt{3} \times Z \times I \times \text{distance}}{1000}$$
$$= \frac{0.6371 \times 1.73 \times 105 \times 3000}{1000}$$
$$= 347.2 \text{ V } (3.2\% \text{ of } 11 \text{ kV})$$

Volt drop is seldom a problem at MV, even for long runs of small conductor size as shown above.

Question 3.2 for course participants: Why would volt drop seldom be a problem at higher voltages?

Prospective fault current

Electric cables are designed to operate below a certain maximum temperature, this being dependent on the conductor material and the type and the thickness of the insulation.

Cable selection for a particular installation must therefore be made on the basis of not exceeding these temperature limits.

Suppose the 400 V distribution board from which a cable is fed has a fault level of 5 MVA. This translates to a fault current of 7.22 kA, and the cable must be capable of carrying this current without damage until the fault is cleared. Fault current ratings for cables are given in the manufacturers' specifications and tables and must be modified by taking into account the fault duration.

The smallest cable that can safely handle the above fault current for a 1 s fault, is a 70 mm^2 copper conductor cable, or a 95 mm^2 aluminum conductor cable. Suppose now that the fault clearance time (including any mechanical delays on the tripping mechanism) is closer to 2 s, then the smallest cable would be a 95 mm^2 copper conductor (11.52/$\sqrt{2}$ = 8.13 kA for 2 s) or a 150 mm^2 aluminum conductor (10.80/$\sqrt{2}$ = 7.64 kA for 2 s).

Likewise, a fault duration less than 1 s will allow the use of smaller conductors than were calculated for the 1 s rating.

Continuing with the example in Section 3.7:

Prospective symmetrical (short-circuit) current:

$$= \frac{250\,000\,000}{\sqrt{3} \times 11\,000}$$

$$= 13.137 \text{ kA}$$

Cable short-circuit current withstand is:

$$I_{SC} = \frac{\text{Cross-section} \times K}{\sqrt{t}}$$

K is 115 for copper conductors, paper insulated
The half second rating is thus:

$$= \frac{35 \times 115}{\sqrt{0.5}}$$

$$= 5.692 \text{ kA}$$

This cable will not survive the prospective short-circuit current. Conductor size required is thus:

$$\frac{x \times 115}{\sqrt{0.5}} = 13.137 \text{ kA}$$

$$\therefore x = 80 \text{ mm}^2$$

The nearest standard size is a 95 mm^2 copper conductor. This has a current rating of 235 A under standard conditions.

Earth fault current for half a second

Lead area for a 95 mm^2 × 3-core PILC cable is 198.2 mm^2.

Cable earth fault current withstand is $= \dfrac{198.2 \times 0.024}{\sqrt{0.5}}$

$$= 6.7 \text{ kA for half a second}$$

$$\text{Required earth fault current, } I_{EF} = \frac{100 \times 10^6}{\sqrt{3} \times 11\,000}$$

$$= 5.25 \text{ kA for half a second}$$

In many cases, the cable conductor size is larger than dictated by the full load current, and is chosen in order to survive the prospective short-circuit current.

The use of large conductors can be avoided by improving the speed of protection (fuses for example) and in the case of earth fault current, by the use of sensitive earth fault protection.

Another example is shown in Figure 3.1.

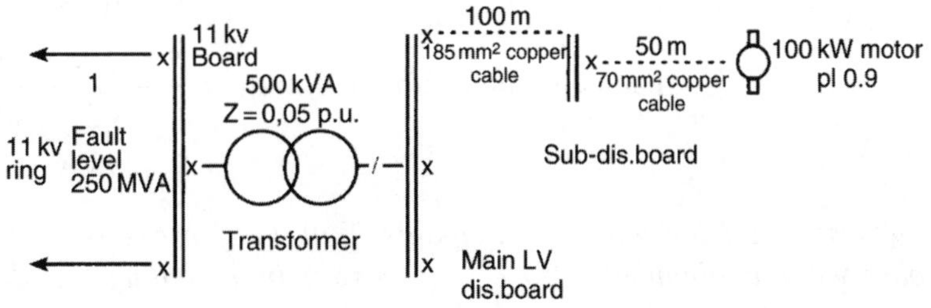

Figure 3.1
Distribution example

We are required to calculate the fault current at the sub-distribution board.
P.U. system:

$$\text{p.u. source impedance} = \frac{\text{Base MVA}}{\text{Fault level MVA}} \tag{1}$$

$$\text{p.u. transformer impedance} = \frac{\text{Base MVA} \times \text{transformers \% impedance}}{\text{Transformers MVA}} \tag{2}$$

$$\text{p.u. cable impedance} = \frac{Z \text{ (ohms)} \times \text{Base MVA}}{V^2 \text{(kV)}} \tag{3}$$

$$\text{p.u. total impedance} = (1) + (2) + (3)$$

$$\text{Fault level} = \frac{\text{Base MVA}}{\text{p.u. impedance}}$$

$$\text{Fault current} = \frac{\text{Fault level (MVA)} \times 1\,000\,000}{\sqrt{3} \times V}$$

Example 1

Using a base of 100 MVA and the per unit method, the impedance of the system at the sub-dist board can be determined:

$$\text{Source: } \frac{100}{250} = 0.4 \text{ p.u.}$$

$$\text{Transformer: } \frac{100}{0.5} \times 0.05 = 10 \text{ p.u.}$$

From Table 3.25, the impedance of 185 mm^2 cable = 0.1445 Ω/km.
Thus the impedance of 100 m = 0.01445 Ω

$$Z.p.u. = \frac{Z\,(\text{ohms}) \times \text{MVA}}{V^2\,(\text{kV})}$$

$$= \frac{0.01\,445 \times 100}{(0.400)^2}$$

$$= 9.03 \text{ p.u.}$$

The fault level at the sub-dist board is then found by dividing the base MVA by the total per unit impedance.

$$\text{Fault level at sub board} = \frac{100}{0.4 + 10 + 9.03}$$

$$\text{Fault current at sub-dist Board} = \frac{5.147 \times 1\,000\,000}{\sqrt{3} \times 400}$$

$$= 7.429 \text{ kA}$$

Question 3.3 for course participants: Will a different answer be obtained if a different base value is obtained? Prove your answer by repeating the above example for a base value of 250 MVA.

Reference to Table 3.25 shows that a 70 mm² copper cable can withstand a short-circuit current of 8.05 kA for 1 s. This is below the potential fault level of the system.

However, the duration of the fault or the time taken by the protective device to operate has to be considered. The circuit supplying the motor would very likely be protected by a 200 A fuse or a circuit breaker. Both these devices would operate well within 1 s, the actual time being read from the curves showing short-circuit current/tripping time relationships supplied by the protective equipment manufacturer. Suppose the fault is cleared after 0.2 s. We need to determine what short-circuit current the cable can withstand for this time. This can be found from the expression:

$$I_{\text{SC}} = \frac{A \times K}{\sqrt{t}}$$

Where:

  $K = 115$ for PVC/copper cables of 1000 V rating

  $K = 143$ for XLPE/copper cables of 1000 V rating

  $K = 76$ for PVC/aluminum (solid or stranded) cables of 1000 V rating

  $K = 92$ for XLPE/aluminum (solid or stranded) cables of 1000

and where:

  $A = $ the conductor cross-sectional area in mm^2

  $t = $ the duration of the fault in seconds

So in our example:

$$I_{\text{SC}} = \frac{70 \times 115}{\sqrt{0.2}} = 18 \text{ kA}$$

This is well in excess of the system potential fault level and below the bursting capacity of the cable so it can be concluded that a 70 mm² cable is suitable for this example. Bursting is not a real threat in the majority of cases where armored cable is used since the armoring gives a measure of reinforcement. However, with larger sizes, in excess of

300 mm², particularly when these cables are unarmored, cognisance should be taken of possible bursting effects.

When the short-circuit current rating for a certain time is known, the formula $E = I^2t$ can also be used to obtain the current rating for a different time. In the above example:

$$I_1^2 t_1 = I_2^2 t_2$$
$$\Rightarrow I_2 = \sqrt{I_1^2 t_1 / t_2}$$
$$I_2 = \sqrt{(8.05^2 \times 1)/0.2}$$
$$= 18 \text{ kA}$$

Note: In electrical protection, engineers usually cater for failure of the primary protective device by providing back-up protection. It makes for good engineering practice to use the tripping time of the back-up device, which should only be slightly longer than that of the primary device in short-circuit conditions, to determine the short-circuit rating of the cable. This then has a built-in safety margin.

Environmental conditions of installation

The data used for determining the current ratings given in this manual are based on calculations according to IEC 287. Ratings for multicore cables are given for a single cable run; where groups of cables run in a common route, the appropriate derating factors are given in the appropriate sections depending on the type of cable being used.

Similarly when the installation conditions differ from standard, the derating factors in the appropriate sections must be used.

A qualitative assessment of the conditions immediately surrounding the cable should be made. Factors such as the need for a fire-retardant sheath, additional mechanical protection, safeguards against chemical attack and corrosion should be considered in this category. These influences affect mainly the external finish of the cable, the armoring and serving.

Cables are sometimes specified with a termite-repellent sheath. It is worth mentioning that no sheath will repel insects since sheath material has to be ingested by the termite to be fatal. Thus, the statement 'cable with a termite repellent sheath' is somewhat erroneous. The PVC sheath is coated with Teflon that makes it difficult bite into the PVC.

In the case of paper, lead cables there are several alternative sheath metals available, pure lead is prone to fatigue when subjected to vibration. An alloy sheath can be supplied which will resist deterioration through fatigue.

The foregoing notes are by no means an exhaustive treatment of the points to consider when choosing a cable for a particular application but are rather guidelines to the salient points that need consideration.

Cables laid directly in the ground

The ratings given are based on a ground thermal resistivity (g) of 1.2 Km/W. The factor (g) varies considerably with differing ground conditions and has a pronounced effect on a cable's current-carrying capacity. The only sure way to determine (g) is to measure it along the cable route. This practice is normally reserved for supertension cables, but there could be other applications where soil thermal resistivity is critical.

Suitability of soils for bedding and backfill

Clay

Clay is a dense, compact material, greasy to the touch when wet and which has a low thermal resistivity even in the fully dried-out condition. Most types of clay however, shrink when drying and can thus not be used as a bedding for the cable. They can be used as backfill and should be consolidated by rolling rather than stamping.

Sand

Sand is a crumbly material with particle grains easily distinguished and gritty to the touch even when wet. Particle sizes larger than 2 mm are known as gravel. Sea sand or sand obtained from a river bed usually consists of spherical particles and has a very high thermal resistivity when dry. Therefore, it should not be used as cable bedding in its pure form.

Some quarried sands and man-made sands as used for making concrete, have irregularly shaped particles of varying size and can be compacted to a high density. These can be used as a bedding material especially when 5–10% clay is added and will have a satisfactorily low thermal resistivity in the dried-out state. Sand/gravel mixes should be used with care as sharp particles can damage the cable serving.

Sand Clay

Sand clay is, as its name implies, a mixture of sand and clay. It is an ideal material for use as bedding and backfill and is best compacted by rolling. It rarely dries out to lower than 6% moisture content.

Loam

Loam can vary in color from reddish brown to dark brown and may contain quantities of organic matter. It crumbles well, even when dry, and can be well compacted to achieve satisfactory values of thermal resistivity. It is very suitable as a bedding material.

Chalk

Chalk is a soft white or gray porous material having a lower thermal resistivity when wet, but dry out to very high values and is unsuitable for use as bedding or backfill in any area where drying out is likely.

Peat

Peat or humus is composed mainly of organic material and is black or dark brown in color. It should not be used as bedding or backfill as dried-out thermal resistivity values of over 4 Km/W are usually obtained. It should be removed and alternative material used for both bedding the cables and backfilling the trench.

Make-up soil

This is a general term for the soil in any area, the level of which has been raised artificially using imported fill which may consist of bricks, concrete, cinders, ash, slag, stones, other refuse or any of the material considered above. This should only be used for backfill and never for cable bedding. If any doubt exists as to suitability, it is best removed completely and a suitable material imported.

Mine sand

Mine sand is thermally very satisfactory, but is highly corrosive and should therefore not be used.

The thermal resistivity of a substance is greatly influenced by the moisture content at a given time (Table 3.1). The higher the quantity of retained moisture, the lower will be the thermal resistivity. A heavily loaded cable will dry out the soil around the cable and cause an increase in (g). This process is cumulative and damage could be done to the cable insulation through overheating.

Material	Thermal Resistivity (g) Km/W
Sandy soil	1.20
Clay	1.60
Chalky soil	1.80
Concrete	0.90
Water-logged ground	0.50
Gravel	1.00

Table 3.1
Typical values of thermal resistivity of substances encountered in cable installations

Impurities such as slag, ash and the like increase the value of (g), as does intense vegetation on the cable route, by drawing moisture from the ground.

Cables installed in air

Multicore cables should be installed with a space of ≥ 0, 3 × overall diameter and single-core cables with a space of ≥ 0, 5 × overall diameter between themselves and the vertical wall or surface supporting them as per IEC 287. If they are installed in direct contact with the wall then the current rating given should be reduced by 5% as a rough guideline provided there is a space of 150 mm or six times the overall diameter of the cable, whichever is the greater between adjacent cables or cable groups in the case of single-core cables. If the installation fails to comply with this requirement then the derating factors in the relevant sections should be applied.

Where the ambient temperature along a route varies, the highest value should be taken to select the cable size.

Question 3.4 for course participants: Why should cables be de-rated when they are installed close to each other or against a wall?

Cables installed in ducts

The air within a pipe or duct will increase the thermal resistance of the heat dissipation path. Consequently, the current rating for a cable run in a duct or pipe is lower than that for an equivalent cable in the ground or in free air. The ratings given can be applied to cables laid in concrete, asbestos, pitch fiber, PVC, earthenware or cast iron pipes which are the more common materials encountered. It should be noted that single-core cables forming a part of an AC system should not be individually installed in cast iron pipes due to the heavy losses incurred by eddy current induction.

Generally, the size of the duct or pipe chosen should depend upon the ease of pulling in, or out, the cable. It should be borne in mind that a larger cable may be required in the future to cater for increased load growth. Common pipe sizes used are 100 mm and 150 mm internal diameters. When groups of cables are run in pipes along the same route, they should be de-rated according to the factors given in the relevant tables.

Composite cable routes

It frequently happens that a cable run is made partly in air, partly direct in the ground and partly in ducts. The latter conditions lead to the lowest rating and it is here that attention must be focused. Very little heat travels longitudinally along the cable, the main dissipation being vertically through the duct wall and surrounding ground. Any rating where the route is part ground, part duct must therefore be treated with care.

Where the length of ducting does not exceed 5 m per 100 m of route length, the cable rating may be assumed to be that for direct burial in the ground.

Intermittent operation

Certain types of loads have an intermittent characteristic where the load is switched on and off before the cable has time to cool completely. Depending upon the load cycle it may be possible to select a smaller cable for intermittent operation than would be the case if the load were continuously applied. When a current in excess of the normal rated current is applied, the heating of the cable will be a correspondingly quicker operation than the cooling.

Generally:

I = Equivalent RMS current
I_n = Current flowing during nth period (including periods of zero current)
t_n = Duration of nth period
N = Number of periods, (includes periods of zero current)

Thus:

$$I = \sqrt{\frac{\sum_{n=1}^{n=N}\left[(I_n)^2 t_n\right]}{\sum_{n=1}^{n=N}(t_n)}}$$

Example:
Suppose a process cycle is as follows:

    150 A for 1 min
    50 A for 2 min
    100 A for 3 min
    0 A for 4 min
    then applying the expression for RMS current

$$I = \sqrt{(57\,500/10)}$$
$$= 75.83\,\text{A}$$

Thus a continuous current of 76 A flowing over the 10 min cycle time would produce the same heating effect as the individual cyclic currents, and the size of cable could be selected based on 76 A.

Solar heating

When cables are installed in direct sunlight, an appreciable heating due to solar absorption takes place. This result in a significant reduction in the cables' current-carrying capacity and for this reason it is strongly recommended that cables be protected from direct sunlight. The maximum intensity of solar radiation measured in Australia varies between 100 and 115 mW/cm^2 depending upon location.

3.8   Transport, handling and installation of electric cables

Cables on wooden drums

- The cable drum is manufactured from carefully selected wood with a low moisture content (typical not more than 15%). If they are required to be treated it shall be done with a preservative or with chromate copper arsenate. Certain types of hardwood do not require treatment.
- Marking on drum flanges should be clear, stencilled or burned into the wood and should include the following information (this will depend on the specific manufacturer, but may be requested by the client):
    - Manufacturer's name or trade mark
    - Rated voltage, rated area, number of cores and specification
    - Length of the cable in meters
    - Year of manufacture
    - Gross mass in kilograms
    - The instruction 'not to be laid flat'
    - Serial number or other identification
    - On each flange an arrow with the words 'roll this way'
    - The standard to which the cable is manufactured.
- Both ends of the cable on the drum should be sealed and the inner end fixed to the flange of the cable drum to prevent loose coiling. The outer end is fixed to the flange as well, for the same reason.
- Cable drums should stand on firm, well-drained surfaces.
- In the past, it was common practice to rotate stored cable drums through 180° to redistribute the impregnation oil through the dielectric. The use of MIND cables has obviated this reason for rotating drums. However, it is still recommended that wooden cable drums that are stored in the open, irrespective of the type of cable contained, should be periodically rotated to avoid the drum timber rotting through rising damp.

Cable transportation

Preparation

- The truck must match the drum size
- Do not overload the trucks
- Cable ends must be sealed, secured and protected
- Use special cable trailers for transportation from depot to site if possible (see Figure 3.2).

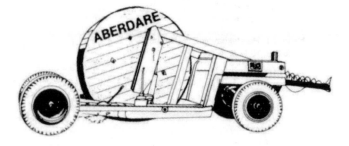

Figure 3.2
Cable trailer

Loading

- Check drums for correct cable and size, serial number, mass and possible damage
- Select correct forklift/crane
- Select correct slings and spindle and check sling condition
- If a crane is to be used, ensure that a spreader is incorporated to prevent damage to drum flanges
- If the drum is to be rolled, observe correct rolling direction by referring to arrows on flanges
- Ensure that the drum bolts are tight
- Ensure that truck surface is clear of obstructions, nails, etc.
- Do not drop drums onto truck loading bed.

Securing

- Secure drums to the truck bed to prevent sliding and rolling, using adequate steel chains and chocks
- Always try to pack drums flange-to-flange
- Do not lay drums flat
- Stop the vehicle during transportation and check that the load is secure.

Off-loading

- Check for damage to cable drums
- Select correct spindle slings for the drum size and mass and ensure that same are in good order, ensure that a spreader is used
- Do not drop drums but lower gently onto firm and relatively level surfaces
- Off-load drums in such a way that they are easily accessible
- If using a fork-truck, ensure that it is of adequate size relative to the task at hand
- Ensure that the fork-truck tines lateral spacing is correct
- Take care that the protruding tines do not damage other equipment or drums
- There are two methods of rolling drums from loading beds if cranes are not available (see Figure 3.3).

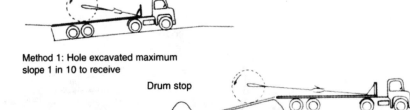

Method 1: Hole excavated maximum slope 1 in 10 to receive

Drum stop

Method 2: Ramp constructed maximum slope 1 in 4.

Figure 3.3
Unloading cable drums

Cable storage

Indoors

- Stack flange-to-flange and preferably not one on top of the other
- Stack so that drums are easily accessible
- Observe fire prevention rules
- Cable ends must be sealed at all times
- Dispatch on 'first in-first out' basis
- Rotate paper-insulated cable drums one complete revolution per annum.

Outdoors

- Drums should be stored on a hard surface at a slight angle and the area should have a drainage system
- Drums should be released on a 'first in-first out' basis
- Cable ends should be sealed at all times
- Stack flange-to-flange but if this is not possible, limit vertical stacking practice to smaller drums only
- Stack in such a way that drums are easily accessible
- Observe fire protection rules
- Cable racks are ideal for storage but take care not to overload
- Cables must be identifiable at all times
- If drums are expected to be stored for a long time they must be specially treated or made of hard wood
- Rotate paper-insulated cable drums one complete revolution per annum.

Mechanical forces on cables during installation

Any cable has a maximum pulling force, which should not be exceeded during installation. The cable construction imposes the limitation on the pulling-in force. When a cable stocking is used, the maximum force can be related to overall cable diameter in mm as follows:

Steel wire armored cables: $F = 0.94 \, d^4 \times 10^{-6}$ kN

Steel tape armored or unarmored cable: $F = 0.39 \, d^4 \times 10^{-6}$ kN

Control and communication cables: $F = 0.26 \, d^4 \times 10^{-6}$ kN

Attempts should be made to limit the pulling force required to a minimum to avoid stretching the outer layers of the cable. This is particularly relevant where control and communication cables are concerned since instances are known where the cores have finished 2–3 m inside the sheath and insufficient overlap at straight joint positions has necessitated relaying some lengths.

An increase in the pulling force is permissible when the cable is laid by means of a pulling eye attached to the conductors. As a rule of thumb, the following forces may be applied to a conductor:

Copper: 4.9×10^{-2} kN/mm²

Aluminum: 2.94×10^{-2} kN/mm²

Then, for example the maximum force that should be applied via a pulling eye to a 70 mm², 3-core copper cable is:

$$70 \times 3 \times 4.9 \times 10^{-2} = 10.29 \text{ kN}$$

Generally when cables are installed using well-oiled rollers and jacks, the following forces can be expected:

Straight route: 15–20% of cable weight
90° bends: 10–20% of cable weight per bend

Example

Calculate whether it would be permissible to pull in a 300 m length of 6.35/11 kV, 185 mm^2 × 3 core, PILC cable with a winch in one go. The route has two 90° bends along the 300 m distance, and well-maintained cable rollers are used.

Taking the worst case initially, from Table 3.27:

$$\text{Cable mass} = 12\ 290 \text{ kg/km} \times 0.3 \text{ km}$$
$$= 3\ 687 \text{ kg}$$
$$\text{Force on cable} = (0.2 \times 3\ 687 \times 9.8) + (2 \times 0.2 \times 3\ 687 \times 9.8)$$
$$= 21.68 \text{ kN}$$

The maximum permissible pulling force (using a cable stocking):
$$F = 0.39 \times 47.42^4 \times 10^{-6} \text{kN}$$
$$= 1.97 \text{ kN}$$

Therefore, the actual force on the cable would exceed the maximum permissible value. This cable may be pulled in by means of a pulling eye attached to the conductors. (Prove this as an exercise.)

Cables laid in open trenches should be left slightly 'snaked' so that any longitudinal expansion or contraction can be accommodated. Similarly, when cables are installed in cleats or on hangers, a slight sag between fixing points is recommended.

Pulling cables through pipes or ducts

When a cable is pulled through a pipe, friction between the cable serving and the pipe material increases the longitudinal force requirements. Representative values for the coefficient of friction (m) between the more common cable servings and pipe materials are given below (Table 3.2).

Serving Material	Pipe Material	μ
PVC	Asbestos	0.65
PVC	Metal (steel)	0.48
PVC	Pitch fiber	0.55
PVC	PVC	0.35
Bitumenized	Asbestos	0.97
Hessian. or	Metal (steel)	0.76
Jute	Pitch fiber	0.86
	PVC	0.55

Table 3.2
Friction coefficients

This information can readily be used to determine the maximum length of cable that can be pulled through a given pipe without exceeding the maximum permissible pulling force. Take for example the 70 mm^2 × 3 core cable previously quoted. If this is a low-voltage cable with a PVC sheath and it is desired to know the maximum length of PVC pipe it can be pulled through then:

Maximum force = 10.29 kN (by pulling eye)

for PVC to PVC, m = 0.35

$$\text{Force} = \mu \times \text{reactive force}$$
$$= \mu \times \text{cable weight}$$
$$= \frac{\mu \times \text{cable mass}}{102}$$

$$\text{Cable mass} = \frac{10.29 \times 102}{0.35}$$
$$= 2\,998 \text{ kg}$$
$$(1 \text{ kN} = 102 \text{ kg})$$

From Table 3.25 the mass of 70 mm^2 × 3 Core copper cable is 3.85 kg/m. Thus the maximum length of cable that can be pulled through a PVC pipe is:

$$\frac{2\,998}{3.85} = 778 \text{ m}$$

If there are any bends in the route then these will create additional loading and reduce the theoretical length of cable that can be installed.

In certain instances when long runs in pipes or ducts are encountered it may be beneficial to grease the cable with petroleum jelly or some other non-aggressive compound to facilitate the pulling-in.

Considerable damage can be done to cable serving at the mouth of a pipe and precautions should be exercised at such points. This point is achieving more importance with the present day trend toward impermeable anti-corrosive sheaths which have to withstand periodic pressure tests. Included among the protective measures that can be adopted are the fitting of a rubber grommet to the mouth of the pipe and inserting a reasonable thickness of rag.

When unarmored cables are pulled into pipes it will be beneficial to ensure that there is no foreign matter present which could cause damage to the sheath before pulling. Pushing a draw rod through the pipe will usually clear any obstruction.

Preparation for cable laying

Planning

When planning a cable route there are several factors to be considered, among the most important of these are:

- Ground thermal resistivity (TR) tests
- Position of joint bays
- Provision to indicate on the 'as laid' drawings, the serial or drum number of the cable installed
- The use of mass or pre-impregnated non-draining paper-insulated cables, XLPE and PVC dielectrics has all but eliminated the need for special precautions when laying cables on steep slopes in shafts.

Drum handling

- Always use the best hoisting equipment available.
- Do not drop drums of cable onto the ground as this not only damages the drum but will damage the cable as well (especially paper-insulated cable).
- It is most important that a minimum of rolling of the drums on the ground be allowed and then only in the direction of the arrows painted on the flanges.
- When rolling a drum of cable, to change direction use two steel plates with grease between them, and by standing one flange on these plates the cable drum may then be swivelled in the desired direction.
- Position the drum prior to cable-pulling so that the cable is pulled from the top of the drum.
- Note that a drum of power cable can weigh up to 10 tons so make sure that adequate cable drum jacks are used, that the spindle is strong enough to hold the drum and that the jacks stand on firm ground and that they hold the spindle horizontal.
- Site the drum at the most convenient place for cable-pulling, usually at the start of a reasonably straight section near the commencement of the trench work.
- Allow for drum braking.

Ground thermal resistivity

- This often governs the rating of a power cable buried directly, as does the temperature of the soil. Losses for cables running at the maximum temperature, at which the dielectric system can faithfully operate for a maximum life of say, 25 years, are considerable, ranging from 15 W/m for normal distribution cables. Cable conductor temperature and the soil surrounding the cable must be able to dissipate this heat effectively or thermal instability (runway) will result. For example an XLPE-insulated 11 kV cable with conductors running at 90 °C could end up with a surface temperature of about 80 °C resulting in drying-out of the soil. Depth of burial plays an important factor here and has been set at 800 mm. Most MV cable current ratings are calculated with ground temperatures at 25 °C at depths of burial of 800 mm.

  LV cables are normally buried at 500 mm. Soil thermal resistivity (the ability of the soil to conduct or dissipate heat) is standard at 1.2 Km/W.
- The actual soil thermal resistivity along the proposed route should be measured by means of an ERA needle probe, but these are outside the scope of this paper; suffice it to say that different soil compositions along the route will have different rates of heat dissipation and could result in 'hot spots'.
- To overcome this, bedding and backfill soils may have to be 'imported'.

Positioning of joint bays

Ensure that there is sufficient working space, consider passing traffic and other obstructions. If it is not possible to position the joint bays at standard cable length distances, remember that the cable can be ordered in specific lengths. Consider drainage for large bays and try to construct the bays prior to cable pulling to prevent any damage to the cable at a later stage.

Recording cable drum serial numbers

Cable drum serial numbers should be recorded on 'as laid' drawings. In the unlikely event of a cable failure in the future, quoting the cable drum serial number will assist the cable manufacturer in his quality control, as this serial number is related to the manufacturing and raw material management in the factory.

Preparing for cable laying

The following 'vital actions' must be observed prior to a cable pull:

- Cable rollers must be placed between 2 and 3 m apart in the trench (depending on size of cable) (see Figure 3.4)
- Ensure that graphite lubricants have been applied where necessary
- Ensure that each member of the pulling gang knows exactly what he is to do and that communication signals between members are clear
- The trench floor must be clear of stones and other obstructions and the cable bedding correctly dispersed.

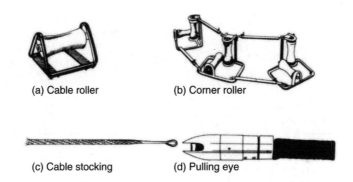

(a) Cable roller   (b) Corner roller

(c) Cable stocking   (d) Pulling eye

Figure 3.4
Cable pulling

Ensure that:

- Cable covers are available at convenient points.
- Any objects that may fall into the trench and damage the cable during the pull and prior to backfilling have been removed.
- If the ambient temperature is below 10 °C or has been so for the past 24 h, the cable on the drum will have to be covered with a tarpaulin and heated with suitable lamps or heaters for at least 24 h under close supervision. Ensure that sufficient ventilation exists, and pay the cable off the drum slowly and carefully. The drum should be lagged with only a few of the bottom lags removed during the heating process (This is necessary only for oil impregnated cables).

Question 3.5 for course participants: Why is this necessary?

- Place the drum at a convenient point prior to the pull on strong jacks and on sound footing (as mentioned earlier) with the arrow on the drum flanges *pointing in the opposite direction* to the rotation when the cable is being pulled.
- Before pulling, cut the inner end of the cable free.
- Remove the drum battens carefully and from the bottom.
- Inspect the cable ends for any sign of leakage (oil impregnated cables). If a leak is suspected, it can be proved by heating the cap until just too hot to touch and if

insulating oil exudes out, the cap should be removed and the extent of the damage assessed, by means of a dielectric test. Cables with extruded dielectrics should be sealed and free from moisture.

- The cable must be paid off from the top of the drum, but take care not to bend it too sharply.
- Cable pulling stockings should be examined and placed over the nose of the cable with care (see Figure 3.4). The pulling rope or wire must be attached to the stocking in such a way that the cable cap will not be damaged during the pull.
- The use of swivels is recommended to prevent twisting of the stocking. The use of stockings is preferable to tying a rope directly to the cable for pulling in.
- For permissible mechanical forces see Section 3.7.
- Bending radius of cables as recommended by the manufacturer should not be exceeded. These are given in Table 3.3.

Cable Type	Bending Radius Up to and Including 11 kV	22 and 33 kV
Paper-insulated cables Single core Multicore	$20 \times d$ $12 \times d$	$25 \times d$ $15 \times d$
PVC-insulated Cables 1000 V Multi- and single core 16–50 sq mm Armored multi- and single core 70 sq mm and greater	$8 \times d$ $10 \times d$	
XLPE-insulated cables Single core Multicore	$17 \times d$ $15 \times d$	$17 \times d$ $15 \times d$

Table 3.3
Recommended bending radii

- Cable should be pulled to its final position in a continuous manner.
- If a winch is being used to pull the cable and unavoidable sharp bends are encountered, a snatch block could be used to assist the pulling tension at the bend.

Once the cable pull is completed, the nose-end of the cable is carefully lifted off the rollers and placed on the bottom of the trench, leaving enough slack to terminate the cable and observing the minimum bending radius. Immediately after cutting, the cable should be suitably sealed on both ends of the cut to prevent the ingress of moisture. Examine the nose cap and make good any damage that may have occurred during pulling.

Bond pulling

These techniques are applied when heavy cables are to be laid or the trench undergoes many changes of direction or very long lengths of cable have to be laid, or a combination of these.

As in the previously mentioned methods of cable pulling, the trench would have been prepared with cable rollers, corner rollers and skid plates. Snatch blocks would have been anchored to the sides of the trench at bends and a winch placed at the far end of the section. At the near end a mobile bond carrier is placed conveniently adjacent to the cable drum, ensuring that its braking system is adequate and that it has rewinding facilities. A steel rope, more than twice the length of the cable to be pulled, is wound onto the bond carrier drum and its end fed over the rollers and through the snatch blocks and then secured to the drum of the winch.

The cable end is manhandled onto the first rollers and tied to the steel rope at intervals of about 2 m (see Figure 3.5). Start the winch and as the nose of the cable arrives at the snatch block untie it from the steel cable, take it around the corner roller and retie on the straight.

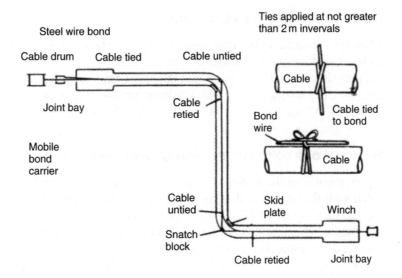

Figure 3.5
Bond pulling

Once the nose has reached the winch end, and allowing the necessary slack, the cable can be untied, the steel rope rewound onto the bond carrier. Further preparation for backfilling may then be commenced.

Backfilling and reinstatement

Once the cables have been laid, and before commencing with backfilling carry out a visual inspection of the installation to ensure that:

- The cables are properly bedded
- Correct spacing between cables if there is more than one in the trench
- Cable entrances at ducts are suitably protected against the possibility of vermin gaining entrance
- Laying and pulling equipment has been removed
- That there is no obvious damage to cable sheaths. Up to 90% of the service failures experienced in any cable system can be avoided if appropriate action is taken at this stage.

Repairs to PVC oversheaths

As mentioned above PVC sheaths damaged 'during pulling' should be repaired, applying the following procedures:

Superficial damage

Generally the local area damaged should be removed, the remaining sheath chamfered for 25 mm at the edges. An EPR self-amalgamating tape is then applied after cleaning the affected area thoroughly with a suitable solvent. The PVC tape should be approximately 30 mm wide and is applied under tension with a 50% overlap continuing up the chamfer until the top is reached plus four layers extending 75 mm beyond the chamfer.

Holes or slits in the PVC sheath

Chamfer the edges of the damaged area for a distance of 30 mm and abrade the area with carborundum strip for a further 20 mm.

Clean the area with a solvent and apply a filling putty followed by a layer of EPR self-amalgamating tape applied at high tension extending 50 mm from the patch, followed by three layers of PVC tape extending 100 mm from the edges of the EPR tape.

Removal of a complete section of oversheath

Upon removal of the damaged ring, chamfer the remaining edges for a distance of 30 mm, clean with the solvent and apply four layers of EPR tape at high tension to 50 mm beyond the chamfer. Apply PVC self-adhesive tape at one-third overlap to a level corresponding to the original oversheath diameter. Five layers of PVC self-adhesive tape are then applied, each one extending 5 mm further along the cable.

The repair is then completed with a resin poultice reinforcement consisting of six layers of ribbon gage or bandage, impregnated and painted with an approved grade of freshly mixed epoxy resin. Allow 12 h to cure.

3.9   Paper-insulated and lead-covered (PILC) 6.35/11 kV cables

The following section generally covers 6.35/11 kV PILC cables.

Notes on impregnating compound

Present day paper-insulated cables are mass impregnated with non-draining compound (MIND). This poly iso butylene compound remains in a solid state at normal operating temperatures and melts at approximately 100 °C.

Compound migration, as was experienced in earlier rosin oil-impregnated cables installed vertically or on inclines, has thus been eliminated by the use of this non-draining compound.

Moisture in paper cables

If a cable is damaged and the lead sheath or end cap is punctured, moisture almost invariably penetrates into the insulation and, if not detected immediately and removed, may cause trouble at a later date. In every such case, therefore, a moisture test should be carried out and the cable cut back until all traces of dampness are removed. The following simple, but reliable test is recommended:

Moisture test

Heat about 1 liter of oil compound (or melted paraffin wax) in a saucepan to a temperature of 150 °C (check by thermometer). Remove individual paper tapes from the cable under test and immerse them in the hot compound. If any moisture is present, it will boil out of the paper and form bubbles or froth, which will rise to the surface of the liquid. If no moisture is present, the hot compound will be undisturbed.

When carrying out the above test do not handle the portion of the paper tapes to be immersed in the compound with bare hands, as moisture from the hands may give rise to false conclusions. As moisture is most likely to travel along the cable under the lead sheath or along the conductors, the papers next to the sheath and conductors are those most likely to contain moisture.

To minimize the penetration of moisture into the cable from the atmosphere or other sources, the cores should be moisture-blocked at each end, by sweating them solid or using solid center ferrules. Table 3.4 shows the design ambient/installation conditions applicable for PILC-insulated 6.35/11 kV grade cables.

Maximum sustained conductor temperature	70 °C
Ground temperature	25 °C
Ambient air temperature (free air-shaded)	30 °C
Ground thermal resistivity	1.2 Km/W
Depth of laying to top of cable or duct	800 mm

Table 3.4
Current rating parameters – base values

Derating factors for non-standard conditions

Tables 3.5–3.9 show the de-rating factors to be used for calculating the current-carrying capacity when the cables are used under non-standard conditions.

Depth of Laying (mm)	Direct in Ground	In Single Way Ducts
800	1.00	1.00
1000	0.98	0.99
1250	0.96	0.97
1500	0.95	0.96
2000	0.92	0.94

Table 3.5
Depth of laying – multicore PILC cables (up to 300 mm²)

Thermal Resistivity (Km/W)	Direct in Ground	In Single Way Ducts
1.0	1.07	1.04
1.5	0.92	0.95
2.0	0.82	0.88
2.5	0.75	0.82

Table 3.6
Ground thermal resistivity (multicore PILC cables)

No. of Cables in Ground	Direct in Ground					In Single Way Ducts			
	Axial Spacing (mm)					Axial Spacing (mm)			
	Touching	150	300	450	600	Touching	300	450	
2	0.80	0.85	0.89	0.90	0.92	0.88	0.91	0.93	
3	0.69	0.75	0.80	0.84	0.86	0.80	0.84	0.87	
4	0.63	0.70	0.77	0.80	0.84	0.75	0.81	0.84	
5	0.57	0.66	0.73	0.78	0.81	0.71	0.77	0.82	
6	0.55	0.63	0.71	0.76	0.80	0.69	0.75	0.80	

Table 3.7
Grouping of PILC cables in horizontal formation at standard soil conditions (multicore cables)

Maximum conductor temperature: 70 °C

Ground Temp (°C)	25	30	35	40	45
De-rating Factor	1.00	0.95	0.9	0.85	0.80

Table 3.8
Ground temperature de-rating factors

Maximum conductor temperature: 70 °C

Air Temperatures (°C)				
25	30	35	40	45
1.10	1.00	0.94	0.87	0.79

Table 3.9
Air temperature de-rating factors

Note: PILC Cables may be grouped in air without derating provided that the cables are installed on cable ladders, and that for:

- *Horizontal formation:* The clearance between cables is not less than 6 × the overall diameter of the largest cable (or 150 mm) whichever is the least.
- *Vertical formation:* The clearance from a supporting wall is greater than 20 mm, and the vertical clearance between cables is greater than 150 mm.

Note: If the number of cables > 4, they are to be installed in a horizontal plane.

Question 3.6 for course participants: Why should more than four cables be installed in a horizontal plane, otherwise de-rating factors would apply?

Short-circuit ratings for PILC cables

Short-circuit ratings do not lend themselves to rigid treatment due to unknown variables, and wherever possible conservative values should be applied.

With the continued growth of power system fault capacity, attention must be given, when selecting a cable, to its short-circuit capacity as well as to the continuous current rating.

Other limiting effects in avoiding damage during subsequent short-circuit conditions are as follows:

- Weakening of joints due to softening of solder at conductor temperatures above 160 °C
- If crimped or welded ferrules and lugs are used, temperatures of 250 °C can be tolerated
- Bursting effects are only of concern with unarmored screened cables larger than 150 mm². Multicore wire armored cables are only likely to burst at currents in excess of 33 kA for cable sizes below 70 mm², in excess of 39 kA for cables below 150 mm² and in excess of 22 kA for cables below 300 mm².

Cable short-circuit ratings are based on the adiabatic performance of the conductors and may thus be regarded as 'internal ratings' which are not affected by external factors as in the case of current ratings. Therefore, no derating factors are needed.

Formula for cable short-circuit current rating:

$$I = \frac{K \times A}{\sqrt{t}} \text{ A}$$

Where

I = short-circuit rating in amps

K = constant combining temperature limits and conductor material properties

A = area of conductor

t = duration of short circuit in seconds

The values of K for copper and aluminum conductors of 6.35/11 kV PILC cables are 154 and 101 A/mm² respectively, for a conductor temperature rise from 70 to 250 °C.

Table 3.27 provides 1 s short-circuit ratings. For other periods, apply the following formula.

$$I_{\text{SC}} = \frac{1 \text{ s short-circuit rating (kA)}}{\sqrt{t}}$$

Earth fault ratings

Some systems make provision for reducing earth fault currents by the inclusion of a neutral electromagnetic coupler (NEC) at the star point of the distribution transformer.

Where this is not the case, the resultant high earth fault current under a fault condition will be carried by the lead sheath and by the galvanized steel wire armor. The bituminized steel tape armor is expected to rust in time and should not be included in any calculation to carry fault current.

The value of K for lead sheaths and galvanized steel wire armor is 24 and 44 A/mm² respectively. The following formula must be applied:

$$I = \frac{K \times A}{\sqrt{t}} \, \text{A}$$

The area of the lead sheath and the armor wires of the cable must be obtained.

3.10 Medium-voltage XLPE-Insulated, PVC-Bedded, SWA, PVC-sheathed cables

(XLPE = Cross-linked polyethylene; SWA = Steel-wired armored)

The following section generally covers 6.35/11 kV XLPE cables.

For XLPE-insulated conductors, continuous conductor temperatures of 90 °C are permissible with overload excursions up to 130 °C for a maximum of 8 h continuous per event, with a maximum total of 125 h per annum. In the case of short circuits, the insulation can withstand conductor temperatures of up to 250 °C for 1 s. Table 3.10 shows the design ambient/installation conditions applicable for XLPE-insulated 6.35/11 kV cables.

Maximum sustained conductor temperature	90 °C
Ground temperature	25 °C
Ambient air temperature (free air-shaded)	30 °C
Ground thermal resistivity	1.2 Km/W
Depth of laying to top of cable or duct	800 mm

Table 3.10
Current ratings are based on these parameters

De-rating factors for non-standard conditions

Tables 3.11–3.15 show the de-rating factors to be used for calculating the current-carrying capacity when these cables are used under non-standard conditions.

Depth of Laying (mm)	Direct in Ground	In Single Way Ducts
500–800	1.00	1.00
850–1000	0.97	0.96
1050–1200	0.95	0.95
1250–1400	0.93	0.95
1450–1600	0.92	0.94

Table 3.11
Depth of laying – multicore XLPE cables (up to 300 mm²)

Thermal Resistivity (Km/W)	Direct in Ground	In Single Way Ducts
0.7	1.23	1.28
1.0	1.08	1.12
1.2	1.00	1.00
1.5	0.90	0.93
2.0	0.80	0.85
2.5	0.72	0.80
3.0	0.66	0.74

Table 3.12
Ground thermal resistivity – multicore XLPE cables (up to 300 mm²)

No.of Cables in Group	Direct in Ground			In Single Way Ducts		
	Axial Spacing (mm)			Axial Spacing (mm)		
	Touching	250	700	Touching	250	700
2	0.79	0.85	0.87	0.87	0.91	0.93
3	0.69	0.75	0.79	0.80	0.86	0.91
4	0.63	0.68	0.75	0.75	0.80	0.87
5	0.58	0.64	0.72	0.72	0.78	0.86
6	0.55	0.60	0.69	0.69	0.74	0.83

Table 3.13
Group of XLPE cables in horizontal formation at standard depths of laying and in standard soil conditions (multicore cables)

Maximum conductor temperature: 90 °C

Ground Temperature (°C)				
25	30	35	40	45
1.0	0.96	0.92	0.88	0.84

Table 3.14
Ground temperature de-rating factors

Maximum conductor temperature (90 °C)

Air Temperatures (°C)				
30	35	40	45	50
1.0	0.95	0.89	0.84	0.78

Table 3.15
Air temperature de-rating factors

Note: Cables may be grouped in air without de-rating provided that the cables are installed on cable ladders, and that for:

- *Horizontal formation*: The clearance between cables is not less than 6 × the overall diameter of the largest cable (or 150 mm) whichever is the least.
- *Vertical formation*: The clearance from a supporting wall is greater than 20 mm, and the vertical clearance between cables is greater than 150 mm.

Note: If the number of cables >4, they are to be installed in a horizontal plane.

Short-circuit ratings for XLPE-insulated 6.35/11 kV cables

Short-circuit ratings do not lend themselves readily to rigid treatment due to unknown variables. Wherever possible conservative values should be applied. As the growth of a power system increases so do the system fault levels. When selecting a cable, attention must be given to its short-circuit capability, as well as to the continuous current rating.

Other limiting effects in avoiding damage during short-circuit conditions are as follows:

- Weakening of joints due to softening of the solder at a conductor temperature of 160 °C and above, although most conductor joining nowadays is done by compression fittings, particularly on XLPE-insulated cables.
- Bucking of the conductors in joint boxes due to longitudinal expansion of cables laid directly in ground.

Cable short-circuit ratings are based on the adiabatic performance of the conductors. This assumes no heat loss from the cable during the period of the fault. No derating factors are necessary with regard to soil temperature, depth of burial, etc.

Ratings are derived from temperature limits as follows:

$$I = \frac{K \times A}{\sqrt{t}} \, \text{A}$$

Where

I = short-circuit rating in amps
K = constant combining temperature limits and conductor material properties
A = area of conductor
t = duration of short circuit in seconds

The value of K for copper and aluminum conductors of 6.35/11 kV XLPE cables is 143 and 92 A/mm^2 respectively, for a conductor temperature rising from 90 to 250 °C

Table 3.28 provides 1 s short-circuit ratings; for other time periods the following formula should be applied.

$$I = \frac{1 \, \text{s short-circuit rating (kA)}}{\sqrt{t}}$$

Earth fault current

Some systems provide for reducing earth fault currents by the inclusion of a neutral electromagnetic coupler (NEC) at the star point of the distribution transformer, to typically 300 A.

Where this is not the case, the resistance of the copper tapes and steel wire armor should be included in the calculation.

Typical 1 s earth fault ratings for XLPE-insulated 6.35/11 kV are shown in Table 3.16.

Cable Size (mm²)	Earth Fault Rating (kA/s)
25	10.4
35	12.2
50	13.1
70	17.6
95	18.7
120	19.7
150	20.8
185	25.0
240	26.8
300	28.6

Table 3.16
Earth fault ratings

3.11 Low-voltage PVC- and XLPE-insulated 600/1000 V power cables

Note on PVC insulation

For PVC insulation continuous conductor temperatures up to 70 °C are permissible. Care must be exercised in matching the cable to the circuit protection. Under short-circuit conditions, a maximum conductor temperature of 160 °C is allowed for a maximum period of 1 s.

Note on XLPE insulation

For XLPE insulation, continuous conductor temperatures up to 90 °C are permissible with excursions of up to 130 °C for a maximum of 8 h continuous per event, with a maximum total of 125 h per annum. Table 3.17 shows the design ambient/installation conditions applicable for XLPE-insulated LV (600/1000 V) cables.

	PVC	XLPE
Maximum sustained conductor temperature	70 °C	90 °C
Ground temperature	25 °C	25 °C
Ambient air temperature (free air shaded)	30 °C	30 °C
Ground thermal resistivity	1.2 Km/W	1.2 Km/W
Depth of laying to top of cable or duct	500 mm	500 mm

Table 3.17
Current rating parameters

De-rating factors for non-standard conditions

Tables 3.18–3.23 show the de-rating factors to be used for calculating the current-carrying capacity when these cables are used under non-standard conditions. These de-rating factors are applicable for both XLPE and PVC-insulated LV cables.

Depth of Laying (mm)	Direct in Ground	In Single Way Ducts
500	1.00	1.00
800	0.97	0.97
1000	0.95	0.96
1250	0.94	0.95
1500	0.93	0.94
2000	0.92	0.93

Table 3.18
De-rating factors for depth of laying – multicore cables (up to 300 mm²)

Thermal Resistivity (Km/W)	Direct in Ground	In Single Way Ducts
1.0	1.08	1.04
1.5	0.93	0.96
2.0	0.83	0.88
2.5	0.78	0.87

Table 3.19
De-rating factors for ground thermal resistivity (multicore cables)

	Direct in Ground					In Single Way Ducts			
	Axial Spacing (mm)					Axial Spacing (mm)			
No.of Cables in Group	Touching	150	300	450	600	Touching	300	450	
2	0.81	0.87	0.91	0.93	0.94	0.90	0.93	0.95	
3	0.70	0.78	0.84	0.87	0.90	0.82	0.87	0.90	
4	0.63	0.74	0.81	0.86	0.89	0.78	0.85	0.89	
5	0.59	0.70	0.78	0.83	0.87	0.75	0.82	0.87	
6	0.55	0.67	0.76	0.82	0.86	0.72	0.81	0.86	

Table 3.20
De-rating factors for grouping of cables in horizontal formation, at standard depths of laying and in standard soil conditions.

Maximum Conductor Temperature (°C)	Ground Temperatures (°C)					
	25	30	35	40	45	50
70 (PVC)	1.00	0.95	0.90	0.85	0.80	0.70
90 (XLPE)	1.00	0.96	0.92	0.88	0.82	0.76

Multicore Cables (up to 300 mm^2)

Table 3.21
Ground temperature de-rating factors.

Maximum Conductor Temperature (°C)	Air Temperatures (°C)			
	30	35	40	45
70 (PVC)	1.00	0.94	0.87	0.79
90 (XLPE)	1.00	0.95	0.89	0.84

Multicore Cables (up to 300 mm^2)

Table 3.22
Air temperature de-rating factors

No. of cables	1	2	3	6	9
Cable touching	1	0.9	0.84	0.80	0.75
Clearance D^* between cables	1	0.95	0.9	0.88	0.85

D^* is the overall diameter of one cable.

Table 3.23
De-rating factors for grouping of multicore cable installed horizontally in air

Note: Cables may be grouped in air without derating, provided that the cables are installed on ladders, and that for:

- *Horizontal formation*: The clearance is greater than 6 × the cable overall diameter (or 150 mm^2, which- ever is the least).
- *Vertical formation*: The clearance from a vertical wall is greater than 20 mm, and the vertical clearance between cables is greater than 150 mm.

Note: If the number of cables > 4, they should be installed in the horizontal plane.

Short-circuit ratings for PVC and for XLPE-insulated 600/1000 V cables

With PVC and with XLPE-insulated cables, care must be taken to limit the conductor temperatures for continuous operation and for short-circuit conditions as indicated in Table 3.24.

Short-circuit ratings are regarded as internal ratings. Their calculation is based on an adiabatic equation and is not affected by external consideration. Due to unknown variables, short-circuit ratings do not lend themselves readily to rigid treatment, so whenever possible conservative values should be applied.

$$I_{SC} = \frac{K \times A}{\sqrt{t}} \ \text{A}$$

Where

I_{SC} = short-circuit rating in amps

K = a constant combining temperature limits and properties of conductor materials

A = area of conductor

t = duration of short circuit in seconds

Insulation Material	Conductor Material	Operating Temperature (°C)	Short Circuit Temperature (°C)	*K* Factor
PVC	Copper	70	160	
PVC	Aluminum	70	160	
XLPE	Copper	90	250	
XLPE	Aluminum	90	250	

Table 3.24
Values of conductor/temperature constant K

Copper conductors

Tables 3.25–3.30 show the electrical and physical properties of different types of cables for use in cable-sizing calculations. Diameters *D1*, *D2* and *d* referred in these tables are as per the legend of Figure 3.6 which shows the typical constructional details of a PVC-insulated 3-core armored cable. The types of cables for which the physical and electrical properties are shown in these tables as follows.

D1 = diameter over bedding
d = diameter of armor wires
D2 = diameter over outer sheath

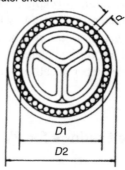

Figure 3.6
Cross-sectional view of a PVC 3-core cable

| | Electrical Properties | | | | | | Physical Properties | | | | | | | |
| | Current Ratings | | | | | | Nominal Diameters | | | | | | Approx. Mass | |
Cable Size (mm²)	Ground (A)	Ducts (A)	Air (A)	Impedance (Ω/km)	Volt Drop (mV/A/m)	1 Short-Circuit Rating (kA)	D1 – 3c (mm)	D1 – 4c (mm)	d – 3c (mm)	d – 4c (mm)	D2 – 3c (mm)	D2 – 4c (mm)	3c (kg/km)	4c (kg/km)
1.5	23	18	18	14.48	25.080	0.17	8.51	9.33	1.25	1.25	14.13	14.95	448	501
2.5	30	24	24	8.87	15.363	0.28	9.61	10.56	1.25	1.25	15.23	16.18	522	597
4	38	31	32	5.52	9.561	0.46	11.40	12.57	1.25	1.25	17.02	18.39	667	762
6	48	39	40	3.69	6.391	0.69	12.58	13.90	1.25	1.25	18.4	19.72	790	910
10	64	52	54	2.19	3.793	1.15	14.59	16.14	1.25	1.25	20.41	21.96	996	1169
16	82	67	72	1.38	2.390	1.84	16.55	19.18	1.25	1.60	22.37	25.92	1295	1768
25	126	101	113	0.8749	1.515	2.87	19.46	21.34	1.60	1.60	26.46	28.34	1838	2196
35	147	120	136	0.6335	1.097	4.02	20.89	23.97	1.60	1.60	27.89	31.17	2215	2732
50	176	144	167	0.4718	0.817	5.75	24.26	28.14	1.60	2.00	31.46	36.54	2871	3893
70	215	175	207	0.3325	0.576	8.05	27.07	31.29	2.00	2.00	35.47	40.09	3617	4837
95	257	210	253	0.2460	0.427	10.92	31.19	35.82	2.00	2.00	39.99	44.62	4901	6115
120	292	239	293	0.2012	0.348	13.80	33.38	38.10	2.00	2.00	42.18	47.40	5720	7269
150	328	269	336	0.1698	0.294	17.25	36.68	42.05	2.00	2.50	45.98	52.65	6908	9250
185	369	303	384	0.1445	0.250	21.27	40.82	46.75	2.50	2.50	51.12	57.45	8690	11039
240	422	348	447	0.1220	0.211	27.60	46.43	53.06	2.50	2.50	57.13	64.16	10767	13726
300	472	397	509	0.1090	0.189	34.50	51.10	58.53	2.50	2.50	62.20	70.13	12950	16544

Table 3.25
3- and 4-core PVC-insulated, 600/1000 V copper cable

Cable Size (mm²)	Electrical Properties						Physical Properties							
	Ground	Ducts	Air	Impedance (Ω/km)	Volt drop (mV/A/m)	1 Short-Circuit Rating (kA)	D1 – 3c	D1 – 4c	d – 3c	d – 4c	D2 – 3c	D2 – 4c	3c (kg/km)	4c (kg/km)
1.5	27	22	22	15.43	26.726	0.21	8.08	8.85	1.25	1.25	13.70	14.47	416	453
2.5	35	30	29	9.45	16.368	0.35	9.18	10.08	1.25	1.25	14.80	15.70	487	521
4	46	39	37	5.88	10.184	0.57	10.06	11.07	1.25	1.25	15.68	16.69	566	650
6	57	49	46	3.93	6.807	0.85	11.25	12.40	1.25	1.25	16.87	18.02	683	778
10	76	67	61	2.34	4.053	1.43	13.25	14.64	1.25	1.25	19.07	20.46	890	1033
16	99	92	80	1.47	2.546	2.28	15.21	17.68	1.25	1.25	21.03	24.42	1191	1544
25	147	119	138	0.9313	1.613	3.57	18.13	19.86	1.60	1.60	25.13	26.86	1693	2018
35	175	142	168	0.6738	1.167	5.00	19.56	22.32	1.60	1.60	26.56	29.52	2025	2511
50	207	169	204	0.5009	0.868	7.15	22.49	25.76	1.60	1.60	29.69	32.96	2606	3242
70	253	207	256	0.3521	0.610	10.01	25.74	29.81	2.00	2.00	32.94	38.21	3323	4503
95	302	248	312	0.2589	0.448	13.58	28.76	33.1	2.00	2.00	37.16	41.90	4442	5650
120	344	282	362	0.2109	0.365	17.16	31.39	35.87	2.00	2.00	40.19	44.67	5335	6731
150	387	318	416	0.1775	0.307	21.45	34.69	40.12	2.50	2.50	43.49	50.42	6403	8708
185	435	359	478	0.1500	0.260	26.45	39.05	44.77	2.50	2.50	49.35	55.07	8184	10343
240	498	413	557	0.1247	0.216	34.32	44.22	50.58	2.50	2.50	54.52	61.68	10073	12932
300	558	471	634	0.1099	0.190	42.90	48.45	55.56	2.50	2.50	58.35	67.16	12076	15575

Table 3.26
3- and 4-core XLPE-insulated 600/1000 V copper cable

Cable Size (mm²)	Copper Conductors					Aluminum Conductors				
	Electrical Properties			Physical Properties		Electrical Properties			Physical Properties	
	Ground Current Rating (A)	Impedance (Ω/km)	1 s Short-Circuit Rating (kA)	Diameter Over Lead (mm)	Approximate Cable Mass (kg/km)	Ground Current Rating (A)	Impedance (Ω/km)	1 s Short-Circuit Rating (kA)	Diameter Over Lead (mm)	Approx. Cable Mass (kg/km)
25	105	0.8779	3.850	31.38	4890	80	1.4421	2.525	31.38	4415
35	130	0.6371	5.390	33.64	5710	100	1.0492	3.535	33.64	5055
50	160	0.4751	7.700	33.49	6020	125	0.7777	5.050	33.49	5195
70	195	0.3365	10.780	36.50	7080	155	0.5423	7.070	36.50	5790
95	235	0.2499	14.630	39.33	8260	185	0.3972	9.595	39.33	6505
120	265	0.2053	18.480	41.73	9440	210	0.3183	12.120	41.73	7225
150	295	0.1739	23.100	44.36	10770	235	0.2640	15.150	44.36	7980
185	335	0.1481	28.490	47.42	12290	265	0.2166	18.685	47.42	8870
240	380	0.1245	36.960	52.14	14480	305	0.1734	24.240	52.14	10050
300	425	0.1106	46.200	56.15	16940	340	0.1472	30.300	56.15	11415

(Short-circuit ratings based on temperature rise of 70°C – 250°C)

Table 3.27
3-core 6.35/11 kV PILC-insulated copper and aluminum cables

Cable Size (mm²)	Copper Conductors					Stranded Aluminum Conductors				
	Current Rating (Ground) (A)	Impedance (Ω/km)	1 s Short-Circuit Rating (kA)	Diameter Overall (mm)	Approx. Cable Mass (kg/km)	Current Rating (ground) (A)	Impedance (Ω/km)	1 s Short-Circuit Rating (kA)	Diameter Overall (mm)	Approx. Cable Mass (kg/km)
35	170	0.6769	5.008	53.9	5600					
50	200	0.5044	7.154	56.4	6225	155	0.8284	4.724	56.4	5 340
70	240	0.3561	10.016	61.7	8050	190	0.5767	6.614	61.7	6 685
95	290	0.2641	13.593	65.9	9240	225	0.4213	8.976	65.9	7 400
120	325	0.2163	17.170	68.9	10330	255	0.3375	11.338	68.9	8 010
150	360	0.1826	21.463	72.4	11695	285	0.2795	14.173	72.4	8 795
185	410	0.1548	26.470	76.5	13165	320	0.2285	17.479	76.5	9 560
240	470	0.1293	34.340	82.6	14740	370	0.1821	22.676	82.6	10 110
300	520	0.1139	42.925	88.9	18050	420	0.1535	28.345	88.9	12 225

Table 3.28

3-core XLPE-insulated 6.35/11 kV copper and aluminum cables

Rated Area (mm2)	Nominal Diameters		Nominal Mass (kg/km)	Impedance (Ω/km)	Current Ratings Phase Ac or Dc		1	Current Ratings Phase Trefoil Formation			3
	D1 (mm)	D2 (mm)			Ground	Air	Volt Drop (mV/A/m)	Ground	Duct	Air	Volt Drop (mV/A/m)
25	5.95	11.91	366	0.8767	129	139	1.75	127	111	109	1.52
35	7.00	12.96	469	0.6356	171	169	1.27	153	132	133	1.10
50	8.15	15.15	632	0.4745	204	207	0.95	180	155	164	0.82
70	9.79	16.57	880	0.3356	254	262	0.67	221	190	207	0.58
95	11.54	19.04	1160	0.2500	308	325	0.50	265	226	256	0.43
120	12.96	20.24	1413	0.2054	353	379	0.41	301	256	298	0.36
150	14.39	22.07	1734	0.1734	402	435	0.35	338	287	341	0.30
185	16.10	24.80	2145	0.1499	461	504	0.30	381	323	396	0.26
240	18.71	27.81	2725	0.1268	545	602	0.25	442	372	473	0.22
300	21.45	30.75	3375	0.1131	627	697	0.23	499	419	550	0.20
400	24.30	34.10	4395	0.1028	735	815	0.21	565	472	640	0.18
500	26.51	37.13	5299	0.0963	856	948	0.19	634	532	732	0.17
630	33.15	43.62	6965	0.0890	996	1108	0.18	718	603	867	0.15
800	37.70	49.00	9118	0.0852	1131	1258	0.17	792	689	984	0.15
1000	42.25	53.45	11050	0.0819	1241	1386	0.16	856	741	1084	0.14

Note: (1) D1 is the diameter over the conductor
(2) D2 is the diameter over the PVC sheath.

Table 3.29
Single-core PVC- insulated cables with stranded copper conductors

| Rated Area | Nominal Diameters | | Nominal Mass (kg/km) | Impedance (Ω/km) | Current Ratings Phase Ac or Dc | | 1 | Current Ratings Phase Trefoil Formation | | | 3 |
	D1(mm)	D2 (mm)			Ground	Air	Volt Drop (mV/A/m)	Ground	Duct	Air	Volt Drop (mV/A/m)
25	5.95	11.81	328	0.9332	169	174	1.866	151	137	137	1.616
35	7.00	12.86	426	0.6760	205	211	1.352	181	164	167	1.171
50	8.15	14.38	567	0.5036	245	257	1.007	213	192	203	0.872
70	9.79	16.22	824	0.3552	302	326	0.71	260	235	257	0.615
95	11.54	17.97	1071	0.2631	366	404	0.526	312	281	318	0.456
120	12.96	19.82	1304	0.2154	422	475	0.431	355	319	372	0.373
150	14.39	21.42	1628	0.1818	480	542	0.363	397	356	426	0.315
185	16.10	23.63	1995	0.1545	554	629	0.309	449	402	494	0.268
240	18.71	26.69	2461	0.1295	656	753	0.259	522	466	594	0.224
300	21.45	30.05	3182	0.1149	766	881	0.229	589	524	692	0.199
400	24.30	33.30	4117	0.1035	902	1045	0.207	668	592	807	0.179
500	26.51	36.33	5032	0.0963	1040	1182	0.192	750	664	925	0.167
630	33.15	42.79	6641	0.0889	1229	1417	0.178	848	746	1094	0.154
800	37.70	48.84	8535	0.0856	1366	1603	0.171	942	823	1254	0.148
1000	42.25	54.21	10676	0.0831	1486	1790	0.166	1025	892	1400	0.144

Note: (1) D1 is the diameter over the conductor
(2) D2 is the diameter over the PVC sheath.

Tablep 3.30
Single-core XLPE-insulated cables with stranded copper conductors

4

Compensation

4.1 Introduction

The term compensation is used to describe the intentional insertion of reactive power devices, capacitive or inductive, into a power network to achieve a desired effect. This may include improved voltage profiles, improved power factor, enhanced stability performance, and improved transmission capacity. The reactive devices are connected either in series or in parallel (shunt).

Figure 4.1 illustrates the flow of power in an electric circuit. The link has an impedance of $R + jX$, and we assume that $V_1 > V_2$ and that V_1 leads V_2. In most power networks, $X >> R$, and reactive power flows from A to B. The direction of reactive power flow can be reversed by making $V_2 > V_1$. The magnitude of reactive power flow is determined by the voltage difference between point A and B. When R is ignored, the reactive power flow, Q is given by

$$Q = \frac{V_2(V_1 - V_2)}{X} \tag{4.1}$$

The ideal situation is when $V_1 = V_2$, and reactive power flow is zero.

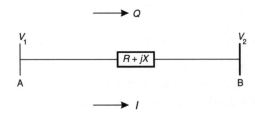

Figure 4.1
Typical power flows in electric circuits

The maximum possible active power transfer, P_{max} is given by

$$P_{max} = \frac{V_1 V_2}{X} \tag{4.2}$$

It is clear from the above formula that active power transfer capacity is improved if V_2 is increased.

4.2   Series capacitors

Series capacitors are utilized to neutralize part of the inductive reactance of a power network. This is illustrated in Figure 4.2. From the phasor diagram in Figure 4.3 we can see that the load voltage is higher when the capacitor is inserted in the circuit.

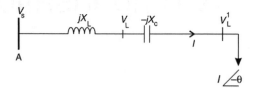

Figure 4.2
Use of series capacitors to neutralize inductor reactors

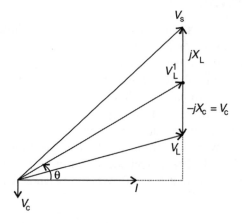

Figure 4.3
Phasor diagram with series capacitor in circuit

Introducing series capacitance in the network reduces the net reactance X, and increases the load voltage, with the result that the circuit's transmission capacity is increased, as can be seen from equation (4.1).

Series capacitors have the following benefits to the network:

- Improved voltage conditions
- Enhanced stability performance
- Controlling reactive power balances
- Aid in load distribution and control of overall transmission losses.

Due to the added transmission capacity, series-capacitor compensation may delay investments in additional overhead lines and transmission equipment, which can have capital investment benefits to the utility company as well as environmental impact advantages.

The first series-capacitor installation in the world was a 33 kV, 1.25 MVAr capacitor bank on the New York Power and Light system in 1928. Many higher-rated systems have since been installed all over the world.

Question 4.1 for course participants: Reducing the net reactance in a system has another direct consequence, which must be taken into account. What is it and what care should be taken?

The installation of series-capacitance in an AC transmission system can result in the phenomenon of subsynchronous resonance (SSR), due to the interaction between the compensated electrical system (in electrical resonance) and a turbine-generator mechanical system (in mechanical resonance). Energy is then exchanged between the electrical and mechanical systems at one or more natural frequencies of the combined system below the synchronous frequency of the system. Increasing mechanical oscillations can occur resulting in eventual mechanical failure of the turbine-generator system.

The following are techniques to counteract SSR:

- *Supplementary excitation control*: The subsynchronous current and/or voltage is detected and the excitation current is modulated using high-gain feedback to vary the generator output voltage, which counters the subsynchronous oscillations.
- *Static filters*: These are tuned to correspond to the power system frequency to filter out the oscillation mode frequencies. They are normally connected in series with each phase of the generator(s).
- *Dynamic filters*: In a manner similar to that of excitation control, the subsynchronous oscillations are detected, a counter emf is generated and injected into the power network through a series transformer to neutralize the unwanted oscillations.
- *Bypassing series capacitors*: To limit transient torque build-up.
- Amortisseur windings on the pole faces of the generator rotors can be employed to improve damping.
- A passive SSR countermeasure scheme involves using three different combinations of inductive and capacitive elements on the three phases. The combinations will have the required equal degree of capacitive compensation in the three phases at the power frequency. At any other frequency, the three combinations will appear as unequal reactance in the three phases. In this manner, asynchronous oscillations will drive unsymmetrical three-phase currents in the generator's armature windings. This creates an mmf with a circular component of a lower magnitude, compared with the corresponding if the currents were symmetrical. The developed interacting electromagnetic torque will be lower.

The following points are worth noting when considering the merits of series capacitors:

- Series capacitors are very effective when the total line reactance is high.
- Series capacitors are effective to compensate for voltage drop and voltage fluctuations.
- Series capacitors are of little value when the reactive power requirements of the load are small.
- In cases where thermal considerations limit the line current, series capacitors are of little value since the reduction in line current associated with them is relatively small.

In the last two points mentioned above, shunt capacitors may be considered instead of series capacitors.

4.3   Shunt capacitors

Shunt capacitors supply capacitive reactive power to the system at the point where they are connected, mainly to counteract the out-of-phase component of current required by an inductive load. They may either be energized continuously or switched on and off during load cycles.

Figure 4.4 illustrates a circuit with shunt capacitor compensation applied at the load side.

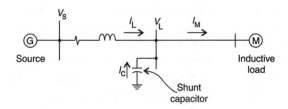

Figure 4.4
Use of shunt capacitors to counteract out-of-phase current component

Referring to the phasor diagram of Figure 4.5, the line current I_L is the sum of the motor load current I_M and the capacitor current I_C.

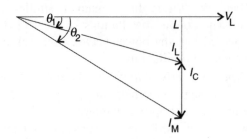

Figure 4.5
Current phasor diagram

It can be seen that the line current is decreased by adding the shunt capacitor. The angle between the load voltage and current is decreased from ϕ_2 to ϕ_1.

Figure 4.6 displays the corresponding voltage phasors. The result of adding the shunt capacitance is to decrease the source voltage from V_{s1} to V_{s2}.

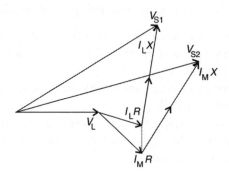

Figure 4.6
Voltage phasor diagram

It can be seen from the above that the application of shunt capacitors in a network with a lagging power factor has the following benefits:

- Increase voltage level at the load
- Improve voltage regulation (if the capacitors are switched in and out of the network correctly)

- Reduce I^2R active power loss and I^2X reactive power loss due to the reduction in current
- Increase power factor
- Decrease kVA (or mVA) loading on the source generators and network to relieve an overload condition or make capacity available for additional load growth
- Reduce demand kVA where power is purchased
- Reduce investment in system facilities per kW of load supplied.

A capacitor starting system may be employed to reduce high inrush currents with the starting of large motors. This aids in maintaining the voltage level in the system. The high inductive component of the starting current is reduced by the addition of capacitance during the starting period only. In this, it differs from applying capacitors for power factor correction.

Note that in determining the amount of shunt capacitance required, some additional capacitive kVAr above that based on initial conditions without capacitors may be required. This is due to the fact that a voltage rise increases the lagging kVAr in the exciting currents of transformers and motors.

Caution

Increased harmonics on the power system and/or a harmonic resonance condition may result with applying capacitors, especially when using harmonic-generating apparatus, such as thyristor controllers. Either a shunt or series resonance condition, or a combination of both, may occur if the resonant point happens to be close to one of the frequencies generated by harmonic sources in the system.

This can result in excessive harmonic currents flowing or harmonic overvoltages, or both, causing possible operation of the capacitor protection equipment (such as fuses), capacitor failure, overheating of other electrical equipment and electrical system interference.

4.4 Shunt reactors

Shunt reactor compensation is usually required under conditions that are the opposite of those requiring shunt capacitor compensation. This is illustrated in Figure 4.7.

Shunt reactors may be installed in the following conditions:

- To compensate for overvoltages occurring at substations served by long lines during low-load periods, as a result of the line's capacitance (Ferranti effect as voltage tip up)
- To compensate for leading power factors at generating plants, resulting in lower transient and steady-state stability limits
- To reduce open-circuit line charging kVA requirements in extra high-voltage (EHV) systems.

The effect of shunt reactance on the current phasor diagram is shown in Figure 4.8.

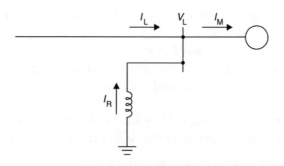

Figure 4.7
Shunt reactor compensation

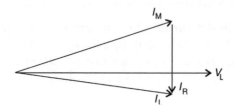

Figure 4.8
Effect of shunt reactors

4.5   Synchronous compensators

A synchronous compensator is a synchronous motor running without a mechanical load. It can absorb or generate reactive power, depending on the level of excitation. When used with a voltage regulator, the motor can run automatically over-excited at high-load current and under-excited at low-load current.

The cost of installation of synchronous compensators is high compared to capacitors, and the electrical losses are considerable relative to capacitors.

Question 4.2 for course participants: Why would the losses of a synchronous motor, utilized as a reactive compensator, be high compared to capacitors?

Synchronous condensers can also be used as dip mitigation devices to support the voltage during drops.

4.6   Static VAR compensators

Static VAR compensators (SVCs) contain shunt capacitors and reactors, which are controlled by thyristors.

They provide solutions to two types of compensation problems normally encountered in practical power systems:

- The first is load compensation, where the requirements usually are to reduce the reactive power demand of large and fluctuating industrial loads, and to balance the real power drawn from the supply lines.
- The second type of compensation is related to voltage support of transmission lines at a certain point in response to disturbances of both load and generation.

The main objectives of dynamic VAR compensation are to increase the stability limit of the power system, to decrease voltage fluctuations during load variations and to limit overvoltages due to large disturbances.

The two fundamental thyristor-controlled reactive power device configurations are:

1. *Thyristor-switched shunt capacitors*: The capacitor bank is split into small capacitor steps and those steps are switched on and off individually. It offers stepwise control, virtually no transients and very little harmonic generation. The average delay for executing a command from the regulator is half a cycle. Figure 4.9 (a) shows this arrangement.
2. *Thyristor-switched shunt reactors*: The fundamental frequency current component through the reactor is controlled by delaying the closing of the thyristor switch with respect to the natural zero crossings of the current. Figure 4.9 (b) illustrates this concept.

Harmonic currents are generated from the phase-angle-controlled reactor. There are two methods to reduce the magnitude of the generated harmonics. The first method consists of splitting the reactor into smaller steps, with only one thyristor-controlled step while the other reactor steps are either on or off.

The second method involves the 12-pulse arrangement, where two identically connected thyristor-controlled reactors are used, one operated from wye-connected secondary winding, and the other from a delta-connected winding of a step-up transformer.

Thyristor-switched reactors are characterized by continuous control, and there is a maximum of one half-cycle delay for executing a command from the regulator.

In many practical applications, a combination of these two thyristor-controlled devices are used, with the SVC consisting of a few steps of thyristor-controlled capacitance and one or two thyristor-controlled reactors, as shown in Figure 4.9 (c).

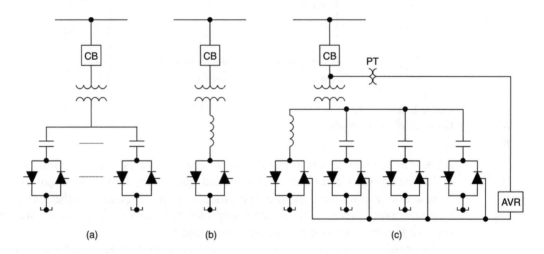

(a)   (b)   (c)

Figure 4.9
Basic static VAR compensator configurations

It is important to note that applying static VAR compensators to series-compensated AC transmission lines results in three distinct resonant modes:

1. Shunt capacitance resonance involves energy exchange between the shunt capacitance (line charging plus any power factor correction and SVCs) and the series inductance of the lines and the generator.
2. Series-line resonance involves energy exchange between the series capacitor and the series inductance of the lines, transformers and generators.
3. Shunt-reactor resonance involves energy exchange between shunt reactors at the intermediate substations of the line and the series capacitors.

Due to the above, it is crucial to represent any compensators in transient electrical simulation programs.

4.7   Distribution applications – shunt capacitors

Shunt capacitors are used more frequently in power distribution systems than any other electrical compensation device. They are used mostly for voltage regulation and power factor correction; hence, these two specific applications will be briefly discussed.

Voltage regulation

Voltage drop can be reduced by the application of a shunt capacitor. A correct selected and located shunt capacitor assures that the voltage at the load will be within the allowable limit at the heavy load condition. However, at light loading, the same capacitor will increase the voltage to above the allowable limit, as illustrated in Figure 4.10.

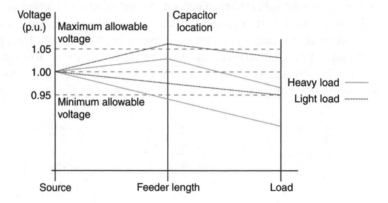

Figure 4.10
Capacitor effect on voltage

The way to avoid this is to use switched capacitor banks. The capacitors are switched in during heavy load conditions and switched out during light load conditions. When the capacitor(s) is switched in, the capacitive current is added to the inductive current, reducing total current, voltage drop and electrical losses. The last is due to reducing reactive power in the system (see next paragraph).

The optimum number, size and location of capacitor banks on a feeder are determined by detailed computer analysis, also taking into consideration minimization of the operation, installation and investment costs. The most important factors that affects the selection is the voltage levels, total loading, distribution factor and power factor of loads.

Power factor correction

Power and power factor

Power in a three-phase distribution system consisting of two components, namely active or real power (P) and reactive power (Q). The complex sum of the two gives apparent power (S). Hence, $S = P + jQ$. This is illustrated in the well-known power triangle in Figure 4.11.

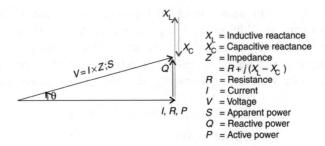

X_L = Inductive reactance
X_C = Capacitive reactance
Z = Impedance
     = $R + j(X_L - X_C)$
R = Resistance
I = Current
V = Voltage
S = Apparent power
Q = Reactive power
P = Active power

Figure 4.11
Power triangle

Apparent power (S) in a three-phase system is calculated by the formula, $S = \sqrt{3}VI$. This is expressed in VA, kVA or MVA, and is the unit used for the ratings of transformers.

Power cables and transformers add reactance to a power network, which is mainly inductive (a lesser amount of capacitance is also added). The inductive and capacitive reactances are frequency dependent (hence are only present in AC systems), oppose each other and are at right angles to the pure (dc) resistance. The net reactance, which is usually inductive, opposes the flow of current, and the power required to overcome this reactance is called reactive power (Q). This is wasted power, which is of no benefit to the user. Reactive power is calculated by the formula, $Q = \sqrt{3}VI\sin\phi$, and is expressed in VAr, kVAr or MVAr.

The presence of reactance in the network causes the voltage and the current phasors at the load to move out of phase by the phase angle ϕ. (The voltage phasor is taken as the reference, and the current phasor is then defined as 'lagging' with respect to a counter-clockwise phase rotation.) The cosine of the phase angle (cos ϕ) is known as the *power factor*.

The longer the cables and the more transformers in a distribution network, the more inductive reactance is present in the network, the greater the phase angle (lower power factor) and the higher the losses. Adding capacitance to an inductive network will serve to decrease the phase angle (ϕ), improve the power factor (cos ϕ) and result in a more efficient distribution network. This is known as *power factor correction*.

When more capacitive than inductive reactance is present in the network, a leading phase angle may result, with the current phasor leading the voltage phasor. A leading phase angle will have detrimental effects on the operation of induction motors, and must be avoided.

Active power is the electrical power available to be utilized usefully, and is expressed in W, kW or MW. This is the unit in which electric motors, lights, heaters, etc. are rated.

Active power is expressed by the formula

$$P = \sqrt{3} \times V \times I \times PF_{(load)}, \quad \text{where } PF = \cos\phi$$

Causes and effects

Any equipment with inductive properties will worsen the power factor, as reactive power is used to overcome the inductive reactance. Therefore, induction motors, transformers and cables will all add to increase ϕ. In addition, variable speed drives that use waveform 'chopping' will also worsen the power factor due to the distortion of the current waveforms, in addition to adding harmonics to the system.

The effect of a low-power factor at the load is that more current is required to achieve the same power output, as can be seen from the power formulae.

The following is an example of the effect of low-power factor:

$$\text{Required active power:} \quad 200 \text{ kW}$$

$$\text{Operating voltage:} \quad 415 \text{ V}$$

$$\text{Case 1:} \quad PF = 0.85$$

$$I = 200\,000 / (1.732 \times 415 \times 0.85)$$

$$= 327 \text{ A}$$

$$\text{Case 2:} \quad PF = 0.55$$

$$I = 200\,000 / (1.732 \times 415 \times 0.55)$$

$$= 506 \text{ A}$$

It can be seen from the above example that power factor has a substantial effect on the magnitude of current flowing in the network.

A sustained low-power factor may have the following effects:

- Overheating of equipment due to the excess current flowing
- Equipment being over-rated to compensate for higher currents flowing due to low-power factor
- Higher electricity consumption when measured in VA
- Necessitate additional investment in system facilities to obtain the required kW
- Lower voltage level at the load (see Section 4.7)
- Increased power losses (resistive and reactive) throughout the system due to higher currents flowing.

Tariff example

Many countries' electricity utilities introduced a penalty charge for customers with low-power factor. Several countries have changed from a kW tariff to a kVA tariff for maximum demand. A low-power factor will have a direct influence on the amount of total kVA a company will use to obtain a certain amount of active kW power (kVA = kW + jkVAr [complex values]). The lower the power factor, the more VARs are consumed.

Therefore, a low-power factor will result in inefficient energy usage and an excessive energy bill. Some utilities penalize consumers if their pf is below 0.96 and others if below 0.80.

The client is then billed for a kVA maximum demand instead of a kW maximum demand.

Let us look at an example:

To simplify the calculation for illustration purposes, we assume average values.

A factory's average monthly maximum demand is 100 MW at a power factor of 0.65.

The power triangle for this situation is shown in Figure 4.12 (a).

For a power factor of 0.65 and real power (P) of 100 MW, the apparent power (S) is 153.846 MVA and reactive power (Q) is 116.913 MVAR [$P = S \cos \phi$; $Q = S \sin \phi$]. As can be noted, the reactive power in the network is of a higher value than the real power!

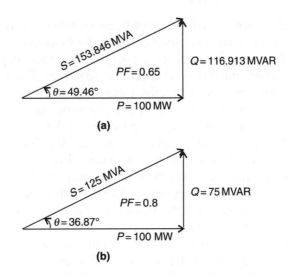

Figure 4.12
Example power triangle

Question 4.3 for course participants: The power factor needs to be higher than what value for the reactive power to be less than the real (active) power?

Referring to figure 4.12 (b), to increase the power factor to 0.8 would need 41 913 kVAR. Therefore, the factory would be charged $0.3757 × 30 × 41 913 = $472 401.42 over a month period, which could be avoided by increasing the power factor to 0.8! This should be more than enough economic justification to install power factor correction (PFC) equipment.

As a very rough approximation, the capital investment for PFC equipment can be taken as $250/kVAR on 11 kV. Therefore, to add −42 MVAR capacitive would mean a capital investment of $10.5 million, which would be paid back in approximately 22 months.

Approving power factor also have added benefits, like reduced active and reactive losses due to the reduction in the magnitude of currents flowing.

4.8 Caution: effect of shunt capacitors on induction motors

Capacitors installed close to induction motors can have the following effects:

- Increasing undesirable torque transients on the rotor
- Self-excitation and capacitive braking.

Torque transients

A short voltage interruption can cause severe torque peaks in the rotor of an induction motor. In the worst case, this can be more than 20 times the full load torque, and in the direction opposite to the rotation of the rotor. Under loaded conditions, this 'reverse torque' can easily cause severe mechanical damage to the rotor shaft.

The presence of capacitance in the system, especially power factor correction capacitors directly on the motor terminals, will worsen the condition substantially.

Explanation of 'reverse torque'

When a running motor is disconnected from the power supply, a trapped or 'frozen' flux is carried round with the rotor. This flux decays, but induces a rotational emf in the stator windings. When the power supply returns while the rotor is still turning, the induced stator emf may have phase-opposition to the supply voltage at the instant of reconnection. This will cause severe transients currents and torque, depending on the magnitude of the induced stator emf still present and the degree of phase-opposition. The transient torque so developed may have a negative (retarding) peak.

The magnitude and direction of the first torque peak are closely dependent on the rotor speed and duration of the interruption, as well as the phase-angle between the induced emf and re-applied voltage. In the worst case the first peak may reach 15 times full load torque (without taking capacitance into consideration).

When capacitors are connected to the stator terminals for e.g. power factor correction, more severe transient effects occur. The capacitors tend to maintain the gap flux when the supply is interrupted, and the stator may build up an overvoltage in spite of the drop in rotor speed. When the supply is reconnected with a phase-opposition to the induced stator emf, very severe current and torque transients will occur. The resultant first transient torque peak may exceed 20 times full load torque.

Normally, the control voltage for the main contactor feeding an induction motor is transformed directly from the main supply voltage. The capacitors connected to the stator terminals will tend to keep the voltage level up during a power interruption, which will in turn keep the contactor coil energized, preventing the contactor from opening. Therefore, when the supply voltage comes back after a short period of time, it is directly applied to the stator, possibly opposing the induced emf already there.

To prevent connected shunt capacitors from worsening the torque transients during voltage interruptions, the capacitor(s) can be disconnected automatically during a severe voltage interruption.

The following is a graphical illustration of the transient torque occurrence (Figure 4.13):

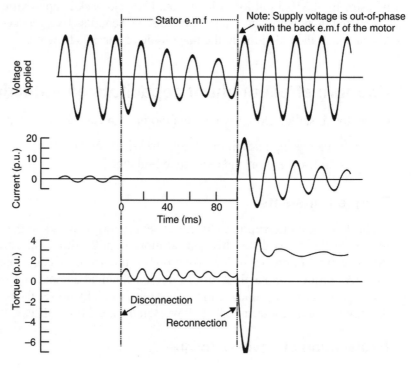

Figure 4.13
Transient torque

Self-excitation and capacitive braking

With the magnetizing reactive power provided by a capacitor bank, provided that the rotor has an adequate remnant field, an induction motor may self-excite upon the loss of stator supply. This results in the motor functioning as an induction generator, and stator overvoltages may occur. The capacitance value of power factor correction capacitors may have to be limited to prevent self-excitation.

Self-excitation may be purposely used under controlled conditions for braking. On disconnection of the stator from the supply and its subsequent connection to a capacitor bank, the stator voltages are self-excited to develop a travelling-wave field, and for a generating mode the 'synchronous' speed of the field must be lower than the speed of rotor rotation. The drive power is drawn from the inertia energy and the speed consequently falls rapidly.

However, for high-inertia loads that stay mechanically connected to the rotor shaft, forced braking may be undesirable and may even lead to rotor shaft damage.

5

Transformer theory

5.1 Transformer theory

Transformer is an essential device in electrical AC power distribution system, which is used to transform AC voltage magnitudes of any value obtained from a source to any desired value for the purpose of distribution and/or consumption. The development of power transformer dates back to 19th century. The main feature of a transformer is its constant *VA* rating whether referred to its primary side or the secondary side. With *VA* being constant (*V* refers to the voltage magnitude and *A* refers to the current magnitude in a transformer winding), it is possible to get a higher *V* with lower *A* or a lower *V* with a higher *A*, by choosing suitable ratio for the transformer.

The major benefit of such a device is its ability to take in the high current produced at relatively low voltage from the electrical generators and transform this power at a higher voltage level with lower current. This ensures that the power generated in the order of several megavolt amperes (MVA) is being transmitted at low current magnitudes in a cable of practical dimensions over very long distances. Today's transmission and distribution systems are heavily dependent upon this technology and transformers are used extensively throughout the world.

The standard transformer

A standard transformer typically consists of a pair of windings, primary and secondary linked together by a magnetic core. The windings can be connected in any of the following types, which in turn decides to which category a transformer belongs. There are two basic types of the transformer viz., (a) shell type and (b) core type. These are illustrated in Figure 5.1.

In Figure 5.1, the windings are the concentric portions and the white solid portions refer to the metallic part of a transformer (laminations), which either surround the windings (shell type) or where the windings surround the core (core type). The windings are normally made of electrolytic grade copper and the metallic parts are made of steel (usually special steel called lamination steel). The metallic part linking the windings is referred to as the magnetic core of a transformer. In these transformers the source will be connected to just one winding only, which is termed as the primary winding. The voltage will be induced magnetically through the magnetic core linking the flux in the core with the other winding (called the secondary winding), since it derives its power from the primary winding.

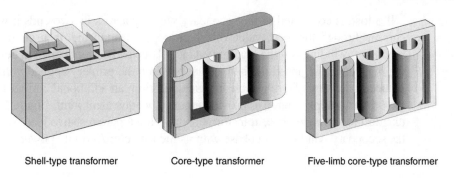

Shell-type transformer   Core-type transformer   Five-limb core-type transformer

Figure 5.1
Transformer types

When an alternating current is applied to the primary winding, an alternating magneto motive force (mmf) is induced into the magnetic core and subsequently sets up an alternating flux passing through its body. This alternating flux in the core that links both the primary and secondary windings induces an electro motive force (emf) in both the windings. In the primary winding this emf is referred to as the back-emf and in an ideal situation, it would oppose the applied voltage to the primary windings such that there would be no flow of current (according to Lenz's law). However the impedance in the windings causes a low current to pass in the primary winding and the current that flows in the primary winding is termed as the transformer magnetizing current. In the open secondary winding, this induced emf is referred to as the secondary open-circuit voltage, which basically refers to the voltage measured between the two ends of the winding. Figure 5.2 depicts the vectorial representation of the voltages, flux and currents that flow in a simple single phase transformer neglecting its impedances. The descriptions of the parameters are identified on the left hand side, which will give a clear idea on the phase angle differences between the voltages, currents and flux in a typical single phase transformer windings and its core.

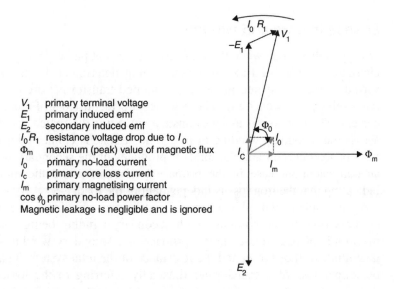

V_1 primary terminal voltage
E_1 primary induced emf
E_2 secondary induced emf
$I_0 R_1$ resistance voltage drop due to I_0
Φ_m maximum (peak) value of magnetic flux
I_0 primary no-load current
I_c primary core loss current
I_m primary magnetising current
$\cos \phi_0$ primary no-load power factor
Magnetic leakage is negligible and is ignored

Figure 5.2
Phasor diagram for a single-phase transformer on open circuit

If a load is connected to the secondary winding across its two ends it will result in flow of current through the load, value of which depends upon the open circuit voltage across the secondary winding and the load impedance. This load current creates a demagnetizing mmf in the magnetic core upsetting the balance between the primary applied voltage and the already induced back-emf. To re-establish this equilibrium an additional current must be drawn from the primary supply source to provide an exactly equivalent mmf. Figure 5.3 shows the phasor diagram for a single-phase transformer supplying a pure resistive load, where the current I_2 in the secondary winding is in phase with its induced emf E_2 (unity power factor).

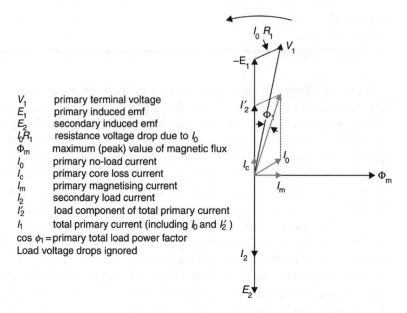

V_1	primary terminal voltage
E_1	primary induced emf
E_2	secondary induced emf
$I_0 R_1$	resistance voltage drop due to I_0
Φ_m	maximum (peak) value of magnetic flux
I_0	primary no-load current
I_c	primary core loss current
I_m	primary magnetising current
I_2	secondary load current
I_2'	load component of total primary current
I_1	total primary current (including I_0 and I_2')

$\cos \phi_1$ = primary total load power factor
Load voltage drops ignored

Figure 5.3
Phasor diagram for a single-phase transformer supplying a unity power factor load

Losses in core and windings

Energy is dissipated when the magnetizing current passes through the core in the form of alternating cycles of flux, the cycles being determined by the system frequency. The various losses dissipated in an open circuited transformer are core loss and no-load loss (also referred as iron loss). These losses are identified in Figure 5.3 above in terms of current. The iron loss is always constant since it is dissipated whether a load is connected across the secondary winding or not. When a load is connected across the secondary, the flow of load current and the mmf. It produces in the secondary winding are balanced by an equivalent increase in the primary load current and the mmf it produces, thereby indicating that the iron loss is independent of the load.

As a fundamental phenomenon, transformer windings also dissipate losses when currents flow in the windings with secondary winding being connected to a load. The magnitude of these losses in the primary and secondary windings will depend upon the magnitude of the current and the resistance of the total system. This is termed as load loss or copper loss of a transformer (basically referring to the losses in copper windings, though modern day transformer windings adopt materials other than copper).

As a fundamental electrical law, the losses in a conductor having resistance value R is $I^2 R$ where I is the current flowing through the conductor. In the same way, the total load loss in a transformer is found to be proportional to the square of the load current since the no-load current produces very negligible resistive loss in the windings.

Leakage reactance

It is not practically possible to ensure that all of the flux produced by the primary winding source is linked to the secondary winding. Due to limitations in the materials used and in the design of transformer cores some flux 'leaks' out of the core and flows through the air – this is termed leakage flux. It is convention to allocate half the leakage flux to both the windings, as it is difficult to quantify exactly the leakage flux contribution to the primary and the secondary windings.

So the transformer can be said to possess leakage reactance due to imperfect transformation between primary and secondary windings. Practically this is expressed as transformer impedance, since transformer windings also have resistance and this impedance is normally expressed as a percentage voltage drop in the transformer at full load current. Expressed mathematically this is:

$$V_z = \%Z = \left[\{I_{FL} \ Z\}/E\right] \times 100$$

Where:

I_{FL} is the full load current (either primary or secondary)

E = open-circuit voltage (either primary or secondary)

$Z = \sqrt{(R^2 + X^2)}$, R and X being the transformer resistance and leakage reactance respectively.

Typical impedance values for a medium-sized power transformer are about 9–10%. Occasionally transformers are deliberately designed to have impedances as high as 22.5%.

Table 5.1 lists typical impedance values for a range of standard transformer ratings, normally found in transmission and distribution systems.

Size MVA	Voltage Rating kV						
MVA	12 kV	33 kV	72.5 kV	145 kV	250 kV	300 kV	420 kV
0.5	4.75%	5.0%	5.5%				
0.8	4.75%	5.0%	6.0%				
5	4.75%	5.0%	6.0%	7.0%			
10	9.0%	9.0%	10.0%	10.0%			
20		10.0%	10.0%	15.0%	13.0%		
30		15.0%	15.0%	12.0%	13.0%		
60		12.0%	12.5%	12.5%	15.0%	15.0%	16.0%
100					16.0%	17.5%	18.0%

Table 5.1
Typical percentage impedance of 50 Hz three-phase transformers

Basic transformer equations

It is observed that the magnetic flux in a transformer core induces voltage in both the primary and secondary windings. Assuming a case where both primary and secondary winding have just one turn each with identical material it is simple logic that there can be no difference between the voltage induced in both of these turns, whether it belongs to the primary or secondary winding. Hence it follows simply that the total voltage induced in each of the windings by the common flux must be proportional to the number of turns used in the respective windings. Thus, this establishes the well-known following relationship in a transformer:

$$E_1/E_2 = N_1/N_2 \tag{5.1}$$

and expressed in terms of power balance:

$$E_1 I_1 = E_2 I_2 \tag{5.2}$$

In terms of current balance,

$$I_1 N_1 = I_2 N_2 \tag{5.3}$$

Where E, I and N are the induced voltages, the currents and number of turns, respectively in the two windings, with suffix 1 denoting the primary winding parameters and suffix 2 denoting the secondary winding parameters. The other known term for IN is ampere turns of a transformer which is a constant factor for a transformer of a particular capacity.

Thus the voltage is transformed in proportion to the number of turns in the respective windings and the currents are in inverse proportion.

Faraday's law establishes that the magnitude of the induced voltage in a winding is proportional to the rate of change of flux linkage associated to the winding. Further Lenz's law proves that the polarity of the induced voltage is such as to *oppose* that flux linkage change if current were allowed to flow, and this is expressed as:

$$e = -N \left(\frac{d\Phi}{dt} \right)$$

In practical terms it can be shown that the voltage induced per turn of a winding is:

$$\frac{E}{N} = K\Phi_m f \tag{5.4}$$

Where, K is a constant, Φ_m is the maximum value of total flux in Webers linking that turn and f is the supply frequency in hertz.

If the voltage is sinusoidal, K is 4.44 and equation (5.4) becomes:

$$\frac{E}{N} = 4.44 f \, \Phi$$

For design calculations, the same can be expressed in terms of volts per turn and flux density in the core:

$$\frac{E}{N} = 4.44 B_m \, A f \times 10^{-6} \tag{5.5}$$

Where
   E/N = volts per turn, which is the same in both the windings (equation (5.1))
   B_m = maximum value of flux density in the core, tesla
   A = net cross-sectional area of the core, mm^2
   f = frequency of supply, Hz
The above equation is basically used to decide the transformer construction parameters, voltages, number of turns, core area, etc.

Equivalent circuit

The above equations prove that there is a simple relationship between the primary and secondary windings in terms of voltage, current, etc. The above equations can be used to relate the impedance parameters to a common side, so that the whole transformer can be converted to a simple inductive circuit with an applied voltage, internal impedance and the current flow being determined by the load connected to the secondary side. In simple terms the circuit will look as shown in Figure 5.4.

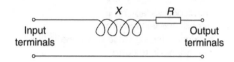

Figure 5.4
Simplified equivalent circuit of leakage impedance of two winding transformer

The transfer of primary impedance on to the secondary is done as follows:

Let Z_S = total impedance of the secondary circuit including leakage and load
characteristics

Z_S' = equivalent value of Z_S when referred to the primary winding

Then

$$I_2' = \left(\frac{N_2}{N_1}\right)(I_2) = \left(\frac{N_2}{N_1}\right)\left(\frac{E_2}{Z_S}\right) \text{ and } E_2 = \left(\frac{N_2}{N_1}\right)(E_1)$$

So

$$I_2' = \left(\frac{N_2}{N_1}\right)^2 \frac{E_1}{Z_S} \tag{5.6}$$

Also,

$$V_1 = E_1 + I_2'Z_1$$

Where

$$E_1 = I_2'Z_S'$$

Therefore

$$I_2' = E_1 Z_S' \tag{5.7}$$

Comparing equations (5.6) and (5.7) it will be seen that $Z_s' = Z_s \ (N1/N2)^2$.

Transformers connected to load

The earlier clauses referred to characteristics of a transformer with a simple resistive load connected to the secondary resulting in secondary current to be in phase with its voltage. However in practical terms, the loads connected to a transformer are inductive in nature. Assume a transformer connected to a load of Z_2 across its secondary. In vectorial representation, Figure 5.5 shows the various parameters of a transformer, which is connected with an inductive load consisting both resistive and reactive components (R and X). The primary parameters are identified with suffix 1 while the secondary winding parameters are suffixed with 2.

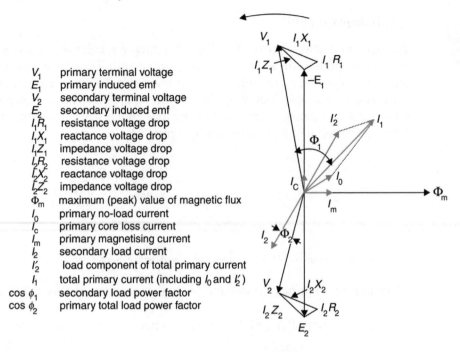

V_1	primary terminal voltage
E_1	primary induced emf
V_2	secondary terminal voltage
E_2	secondary induced emf
I_1R_1	resistance voltage drop
I_1X_1	reactance voltage drop
I_1Z_1	impedance voltage drop
I_2R_2	resistance voltage drop
I_2X_2	reactance voltage drop
I_2Z_2	impedance voltage drop
Φ_m	maximum (peak) value of magnetic flux
I_0	primary no-load current
I_c	primary core loss current
I_m	primary magnetising current
I_2	secondary load current
I'_2	load component of total primary current
I_1	total primary current (including I_0 and I'_2)
$\cos\phi_1$	secondary load power factor
$\cos\phi_2$	primary total load power factor

Figure 5.5

Phasor diagram for a single-phase transformer supplying an inductive load of lagging power factor cos Φ_2.
Voltage drops divided between primary and secondary sides

It is to be noted that the basic phasor diagram still looks the same except for the lagging power factor introduced in the currents of both primary and secondary windings. In fact, the sum total effect is a reduction in the secondary terminal voltage. The drops due to primary resistance and leakage reactance can be converted in terms of the secondary voltage using the transformer equations. Figure 5.6 shows the phasor conditions where the resistance and leakage reactance drops are shown as occurring on the secondary side with all parameters transferred to secondary side adopting the transformer equations.

What happens when the load on the secondary is capacitive in nature? In fact, there will be a rise in the voltage as can be seen in Figure 5.7. This happens when a leading current passes through an inductive reactance (which is the transformer itself).

Preceding phasor diagrams have illustrated the relationship between the various voltages and currents and the phase angles between them for a single-phase transformer. These phasor diagrams are applicable for polyphase transformers too, except that there will be an angular difference between the parameters (phase difference of 120° between the phasewise parameters in a three-phase transformer). In all other respects, the individual phases emulate almost same type of characteristics as single-phase ones.

Rated quantities

As described in the initial paragraph, the transformer rating is specified in voltamperes indicating the amperes *A* can handle at a rated voltage *V*. Depending on the magnitude of power being transferred or handled, the transformer rating or output can be expressed in MVA or in kilovoltamperes (kVA), and are determined by the following equations:

Single-phase transformers

Output = $E_2 \times I_2$ in VA

or $\quad = (4.44\, f\, \Phi_m\, N) \times I$

with the multiplier 10^{-3} for kVA and 10^{-6} for MVA.

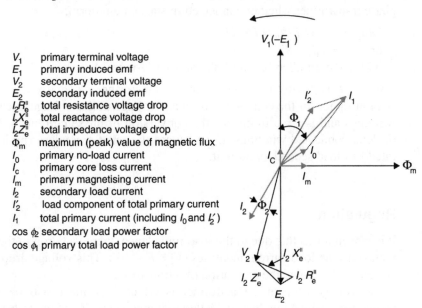

V_1	primary terminal voltage
E_1	primary induced emf
V_2	secondary terminal voltage
E_2	secondary induced emf
$I_2 R_e''$	total resistance voltage drop
$I_2 X_e''$	total reactance voltage drop
$I_2 Z_e''$	total impedance voltage drop
Φ_m	maximum (peak) value of magnetic flux
I_0	primary no-load current
I_c	primary core loss current
I_m	primary magnetising current
I_2	secondary load current
I_2'	load component of total primary current
I_1	total primary current (including I_0 and I_2')
$\cos \phi_2$	secondary load power factor
$\cos \phi_1$	primary total load power factor

Figure 5.6
Phasor diagram for a single-phase transformer supplying an inductive load of lagging power factor cos Φ_2. Voltage drops transferred to secondary side

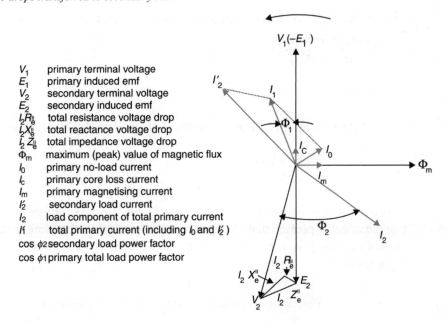

V_1	primary terminal voltage
E_1	primary induced emf
V_2	secondary terminal voltage
E_2	secondary induced emf
$I_2 R_e''$	total resistance voltage drop
$I_2 X_e''$	total reactance voltage drop
$I_2 Z_e''$	total impedance voltage drop
Φ_m	maximum (peak) value of magnetic flux
I_0	primary no-load current
I_c	primary core loss current
I_m	primary magnetising current
I_2'	secondary load current
I_2	load component of total primary current
I_1	total primary current (including I_0 and I_2')
$\cos \phi_2$	secondary load power factor
$\cos \phi_1$	primary total load power factor

Figure 5.7
Phasor diagram for a single-phase transformer supplying a capacitive load of leading power factor cos Φ_2. Voltage drops on primary are transferred to secondary side

Three-phase transformers

A three phase transformer basically consists of three windings and is used to transmit the three-phase power produced by a generator. The three-phase transformer also has primary and secondary windings, same as a single-phase transformer, but the windings of each phase are separately wound and brought out for both primary and secondary with the primary terminals connected to the three-phase source. The three windings of the primary and secondary can be connected in the form of either star or delta. However in a three-phase transformer whether connected in star or delta form,

$$\text{Output} = 4.44 f\, \Phi_m\, N\, I \times \sqrt{3}$$
$$\text{or} \quad = E_2 I_2 \times \sqrt{3}$$

with the multiplier 10^{-3} for kVA and 10^{-6} for MVA.

Where I_2 is the full-load current in the transformer secondary winding. The constant $\sqrt{3}$ is a multiplier for the conversion of phase voltage to line voltage in case of star-connected windings and is a multiplier for the conversion of phase currents to line currents in case of delta-connected windings. E_2 and I_2 are the rated secondary (no-load) voltage and the rated full-load secondary current.

Regulation

It has been noted that due to the load impedance, there is a drop in the secondary terminal voltage as the load current increases ($V_2 = E_2 - I_2 Z$). This voltage drop varies with the load current and is called the regulation of a transformer.

The voltage drop in a transformer is effected by the resistive (R) and reactive (X) components which together are called the impedance (Z) of the transformer.

The copper loss in a transformer is due to the resistive component when a load current passes through the windings. Hence,

V_R = percentage resistance voltage at full load (voltage drop due to resistive component)

$$= \frac{\text{copper loss} \times 100}{\text{Rated kVA}}$$

Similarly,

V_X = percentage reactance voltage (voltage drop due to reactive component)

$$= \left(\frac{I_2 X'_e}{V_2} \right) \times 100$$

Based on the above equations and referring to the phasor diagrams, it can be mathematically proven that the approximate percentage regulation at a particular load current equalling 'a' times full load current of the transformer with a power factor of cos Φ_2 is given by:

$$\text{Percentage regulation} = a \left(V_R \cos \Phi_2 + V_X \sin \Phi_2 \right)$$
$$+ \frac{a^2}{200} \left(V_X \cos \Phi_2 - V_R \sin \Phi_2 \right)^2 \tag{5.8}$$

The percentage regulation that occurs at the secondary terminals, at unity power factor load ($\cos \Phi_2 = 1$ and $\sin \Phi_2 = 0$ may be calculated by simplifying the above equations as below:

$$\frac{\text{Copper loss} \times 100}{\text{output}} + \frac{(\text{percentage reactance})^2}{200}$$

This value is always positive and indicates a voltage drop with load.

For transformers having low reactance values up to about 4%, this can simplify to:

$$\text{Percentage regulation} = a(V_R \cos \Phi_2 + V_X \sin \Phi_2) \tag{5.9}$$

For transformers with high reactance values about 20% and above:

$$\text{Percentage regulation} = a\left(V_R \cos \Phi_2 + V_X \sin \Phi_2\right)$$
$$+ \frac{a^2}{2 \times 10^2}\left(V_X \cos\Phi_2 - V_R \sin \Phi_2\right)^2$$
$$+ \frac{a^4}{8 \times 10^6}\left(V_X \cos \Phi_2 - V_R \sin \Phi_2\right)^4 \tag{5.10}$$

It had been noted that the secondary voltage will get increased by increasing its number of turns ratio with respect to the primary. This characteristic is used to offset the regulation in a loaded transformer. The transformer is provided with additional winding with suitable taps for external connection. Normally this provision is made on the high-voltage side to keep the current to be controlled at a lower value. Whenever there is a voltage drop, the taps are changed to a lower position to get a proportional increase in the turns ratio between primary and secondary so that a higher voltage can be obtained on the secondary side.

On-load tap changers

On-load tap changers (Figure 5.8) are very necessary to maintain a constant voltage on the LV terminals of the transformer for varying load conditions where the voltage had to be maintained constant without interrupting the power supply.

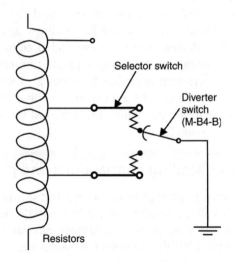

Figure 5.8
On-load tap changer

This is achieved by providing taps as shown below on the HV winding. The diverter switch is controlled by a motor which ensures that the switch is stepped up or down from one tap to the next taps until the desired secondary voltage is obtained.

The ranges are normally chosen depending upon the input supply conditions also. A typical range of taps could be +15% to –5% giving an overall range of 20% as shown in Figure 5.9.

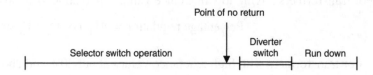

Figure 5.9
Tap changer range of operations

The tap changer is usually mounted on a separate tank to the main tank.

The tap changing can also be done by disconnecting the source (or load) by using off-circuit tap switch, which necessarily requires an interruption in power supply to the load. The off-circuit tap switch is manually changed from one step to another depending on the input conditions. It is a well-accepted practice to provide off-circuit taps from –5% to +5% in steps of 2.5% (total five steps).

The on-load changer which is comparatively costlier is adopted in case of larger and important transformers mostly in continuous process industries, whereas off-circuit change is adopted in the case of almost all smaller distribution or auxiliary transformers.

5.2 Transformer design applications

In shell-type transformers, the flux-return paths of the core are external to, and enclose the windings (refer Figure 5.1 for a three-phase shell-type transformer). The shell-type arrangement provides intrinsically better magnetic shielding that is well suited for supplying power at low voltages and heavy currents. This design is widely used in arc furnace transformers.

The core-type designs are extensively used throughout the world for other types of transformers. These have concentric windings around the various limbs as shown in Figure 5.1. This design possesses top and bottom yokes of cross-section equal to the wound limbs, thereby not requiring separate flux-return paths. In the event of very large transformers, where transportation becomes a problem, the cross-section of the top and bottom yokes are reduced to almost 50% of that of the wound limbs. This necessitates additional flux-return paths which are achieved by adopting five-limb transformer design illustrated in the same figure.

Transformer winding connections and phasor relationships

The most common arrangement for any transformer is to have only two windings per phase, a high-voltage winding and a low-voltage winding (primary and secondary). As already noticed, a three-phase transformer can have its windings connected either in star or in delta with 120° displacement between each phase supply (see Figures 5.10 and 5.11). The following paragraphs explain the different characteristics of transformers depending upon the connection method adopted.

Depending on the method chosen for the primary and the secondary connections, a phase-shift can take place through the transformer from primary to secondary (Figure 5.12).

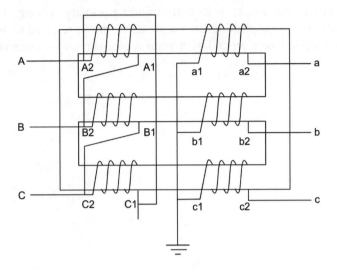

Figure 5.10
Physical connection of delta (D) or star (Y) configuration

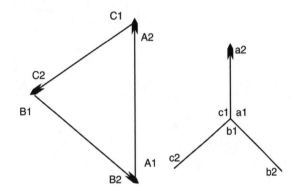

Figure 5.11
Vectorial representation of delta and star configuration

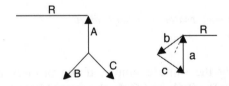

Figure 5.12
30° phase shift of transformer

Figure 5.13 shows the consolidated phase shift phenomenon of a two-winding three-phase transformer, when one winding is delta connected and the other is star connected (A-B-C refer to primary terminals and a-b-c refer to secondary terminals).

In the example shown, the phase shift of secondary winding is 30° with respect to the corresponding primary winding. i.e. the a-n voltage lags A-N by 30°, b-n voltage lags

B-N by 30° and so on. The vector diagram is considered analogous to a clock whose hour hands also rotates 360° same as the full cycle of a vector. Considering a clock with 12-h positions, each hour position corresponds to 30° shift when you start looking after the 12 o'clock position. Hence the above connections achieve an equivalent of 1 o'clock position with respect to its corresponding primary voltage. This connection is referred to as the 1 o'clock position. In simple terms this is referred to as Dyn1 (small n represents a connection where secondary neutral is brought out for external connection).

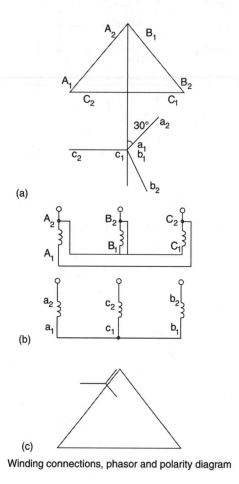

Winding connections, phasor and polarity diagram

Figure 5.13
Winding connections, phasor and polarity diagram

Considering the same example, if the primary delta connection had been made by connecting A_1B_2, B_1C_2 and C_1A_2 this would have resulted in a phase displacement of 30° anticlockwise on the secondary side, i.e. the 11 o'clock position. This type of connection is termed as Dy11 (or Dyn11).

In a similar way the transformer winding connections can be expressed in different ways. For example:

YNd1 = HV winding connected in star with primary neutral (capital N) brought out.

LV winding connected in delta.
Phase shift 30° from 12 to 1o'clock.

Table 5.2 lists the various group numbers that transformer phasors can be categorized, representing the induced emfs and the counter clockwise direction of rotations.

Group Number	Phase Displacement	Clock Hour Number
I	0°	0
II	180°	6
III	−30°	1
IV	+30°	11

Table 5.2
Group numbers

Inter-phase connections of the HV and LV windings are indicated in Table 5.3.

	Winding Connection	Designation
High voltage	Delta	D
	Star	Y
	Interconnected star	Z
Low voltage	Delta	d
	Star	y
	Interconnected star	z

Table 5.3
Winding connection destinations

Winding polarity

International standards define the polarity of the HV and LV windings sharing the same magnetic circuit as follows.

If the core flux induces an instantaneous emf from a low-number terminal to a high-number terminal in one winding, then the direction of induced emf in all other windings linked by that flux will also be from a low-number terminal to a high-number terminal (A1 to A2 and a1 to a2 in Figure 5.14).

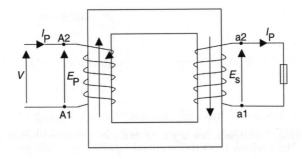

Figure 5.14
Principle of operation of a transformer

From the laws of induction, it will be seen that the current flow in the windings is in the opposite direction.

Figure 5.15 illustrates the testing connections and method to determine the polarity of a transformer.

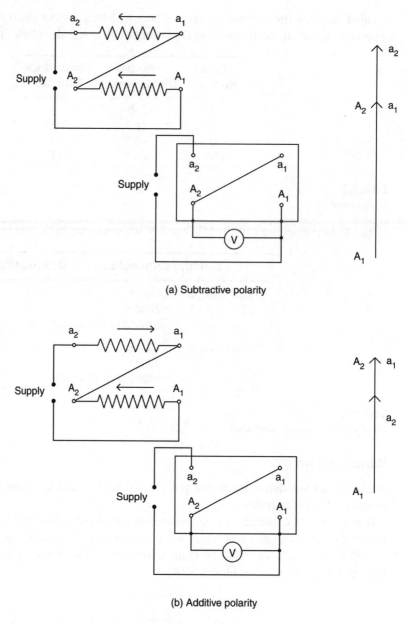

(a) Subtractive polarity

(b) Additive polarity

Figure 5.15
Test connections for determining single-phase transformer winding polarity

The connections basically require interconnecting the phase terminals of primary and secondary windings, applying voltage to one set of winding and measuring the voltage across the various terminals caused by the induction phenomenon. As is evident from the diagrams, if the voltage measured across A_1–A_2 is less than the voltage measures across A_1–a_2 then the polarity is said to be subtractive, and if it is greater than, then the polarity is additive.

Figure 5.16 illustrates the test connections for a three-phase star/star-connected transformer with subtractive polarity.

Polarity forms an important requirement for connecting the current transformers in the transformer windings to ensure that the current is sensed in the correct direction of flow before deciding the fault conditions for tripping purposes. Figure 5.17 shows how the phase relationships and the polarities can undergo changes depending upon the phase sequence followed in the primary connections and changing the star point on the secondary side.

These changes can have different effects on the relays being connected and their operation.

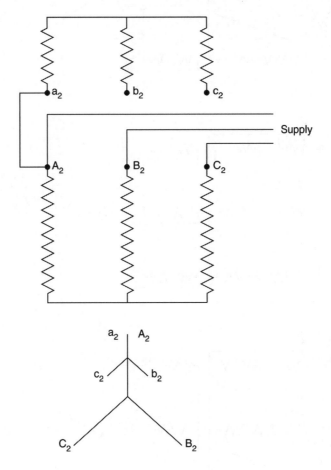

Figure 5.16
Test connections for determining three-phase transformer winding polarity

Transformer grounding

Earthing is a major requirement in most electrical systems to limit the fault currents, and transformers being the prime source of power distribution, the earthing of transformers takes utmost priority in a system. It is obvious that the transformers shall have a neutral reference point for the three phases so that this point can be connected to earth. As a consequence, it is convenient to have most three-phase transformer windings in a star-connection that provides a neutral that could be connected to earth. This connection to earth could be direct or through a current limiting resistor, or other similar devices. Earthing of system neutrals primarily falls in to three categories:

1. Solidly earthed neutral
2. Earthing via an impedance
3. Isolated neutral (unearthed).

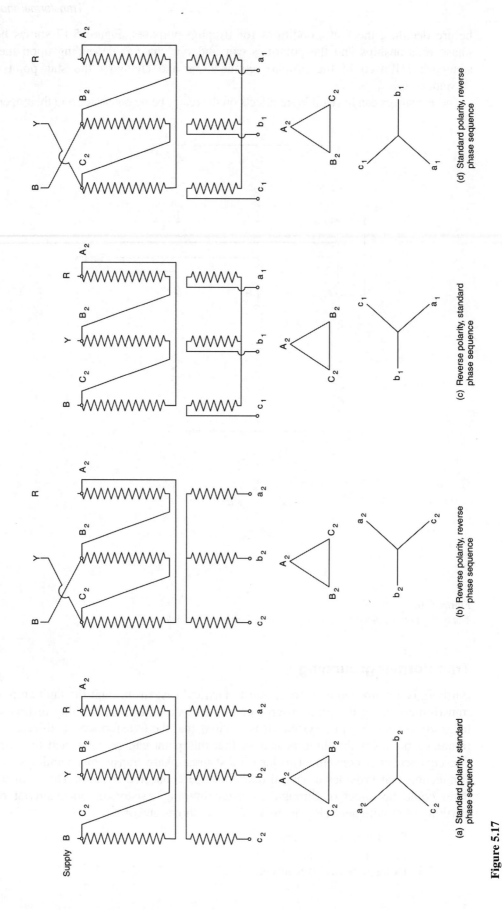

Figure 5.17

Diagrams showing four examples of a three-phase delta/star-connected transformer having different polarity and phase sequence

The third alternative is rarely seen due to the problems and disadvantages associated with an isolated earth. The method of earthing low-voltage and high-voltage electrical systems differ. The definitions of high voltage and low voltages would differ from country to country and even region to region, but in practice, usually any system over 50 V but not exceeding 1000 V is considered a low-voltage system. The low-voltage systems' neutrals are required to be solidly connected to earth and no impedance shall be inserted in any connection with earth, except in cases where they are required for operation of switchgear, etc.

Some of the advantages of connecting high-voltage systems to earth are:

- Earth faults effectively become short circuits from line to neutral thereby protecting equipment from serious damage occurring from high-voltage oscillations.
- An earthed neutral allows rapid operation of protection the instant earth faults occur, working in conjunction with sensitive fault protection devices, to isolate the faulty sections quickly.
- Solid earthing of neutrals ensures that the voltage of any live conductor never exceeds the voltage between the line and neutral, effectively maintaining the neutral at zero volt potential.
- Earthed neutrals provide added protection against capacitance effects on overhead lines that are subject to induced static charge from adjacent charged clouds, dust, sleet, fog, rain and changes in altitude of the lines.
- The HV winding can have graded installation; thus making the transformer smaller and cheaper.

The only disadvantage of connecting a high-voltage system to earth is that it introduces susceptibility to earth faults. This is particularly inconvenient for long overhead transmission lines in lightning prone areas, but as this phenomenon is only transient in nature, and is cleared immediately following a trip, this disadvantage is far outweighed by the advantages listed above.

An earth fault condition is also a reason for flow of third harmonic currents in an electrical system and higher magnitude third harmonics are undesirable. However the third harmonics can be eliminated if a separate path could be created for the flow of third harmonic currents. Hence to provide a path for the third harmonic currents in order to eliminate or attenuate the third harmonic voltages, it is also desirable to have a delta-connected three-phase system. Considering the need for grounding and the need to eliminate third harmonics, in most three-phase step down transformers it is convenient to have the high-voltage (HV) winding delta-connected and the low-voltage (LV) winding star-connected, with the neutral earthed.

Different methods of earthing neutrals

It is important that the neutral of a power system be earthed; otherwise, this could 'float' all over with respect to true ground, thereby stressing the insulation above its design capability.

On HV systems (i.e. 6 kV and above) it is common practice to effectively earth the neutral by means of a solid bar of copper (Figure 5.18).

This has the advantage that when an earth fault occurs on one phase, the two healthy phases remain at phase-to-neutral voltage above earth. This allows insulation of the transformer windings to be graded toward the neutral point, resulting in a significant saving in cost. In addition, all other primary plant need only have phase-to-neutral insulation and surge arrestors in particular need only be rated for 80% line-to-line voltage. This provides an enormous saving in capital expenditure and explains why utilities HV system is invariably solidly earthed.

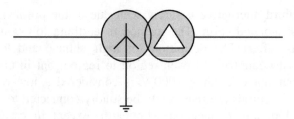

Figure 5.18
Earthing of the neutral

The disadvantage is that when an earth fault occurs, an extremely high current flows (approximately equal to three-phase fault current), stressing the HV windings both electromagnetically and thermally, the forces and heat being proportional to the current squared.

Earthing of the LV system can be achieved as in Figure 5.19.

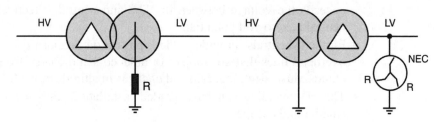

Figure 5.19
Earthing of the LV system

It will be noted that the LV system is impedance and/or resistance earthed. This allows the earth fault current to be controlled to manageable levels, normally of the order of the transformer full load current, typically 300 A.

Here, the transformer does not get a shock on the occurrence of each earth fault, however, the phase conductors now rise to line potential above earth during the period of the earth fault (Figure 5.20). It is recommended to use neutral earthing resistor (NEC/R) to limit severe overvoltages in the healthy phases of a single phase faulted system by damping the oscillations. The idea is supported and encouraged by SABS 0200 and IEEE standard 142.

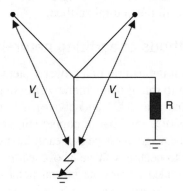

Figure 5.20
Phasor diagram illustrating phase conductors rising to phase voltage on fault

Phase-to-earth insulation of all items of primary plant must therefore withstand line-to-line voltage.

There is a case where the secondary of a transformer is delta connected without neutral. These types of transformers can be connected in the following way (Figure 5.21) as shown in Figure 5.13 for earthing purposes.

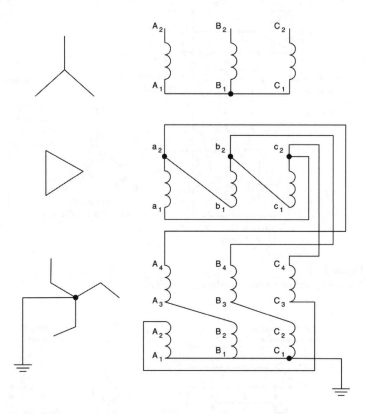

Figure 5.21
Transformer with delta secondary and interconnected star earthing transformer with a neutral connected to earth

The criterion that is to be considered is to decide whether to have a solid connection of the system neutral to earth or to introduce impedance in the circuit to earth. The value of the impedance must be such that the total current flow to earth is limited thereby limiting the amount of damage caused at the point of the earth fault. In keeping with this, Figures 5.22 and 5.23 illustrate two different methods of providing an earthed neutral to a star/delta transformer. Each offers two possibilities of providing earthing resistors and an artificial neutral for the delta windings that normally do not possess one.

In one case only one resistor is required that is designed to carry the total fault current to earth, while being insulated for the total phase voltage of the system. On the other hand, in the event of a fault, the neutral of the transformer will rise above the earth potential, equal to the potential drop across the earthing resistor, and the transformer windings will have to be insulated for the full line voltage above earth.

Alternatively in order to avoid sudden high-voltage changes in the insulated windings which is the most vulnerable component of the equipment, the second option should be chosen of installing suitably sized resistors between the lines and the earthing transformer rather than between the neutral and earth. In doing so the purpose of current limiting is achieved and in addition the neutral of the earthing transformer remains at earth potential and the windings are not subjected to any high voltages.

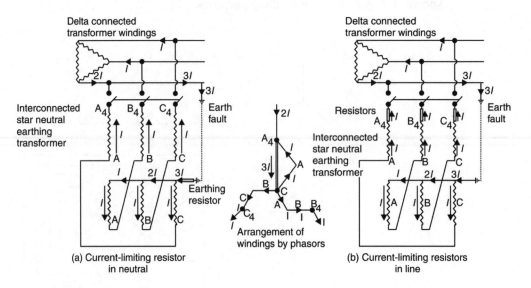

Figure 5.22
Interconnected star neutral earthing transformer

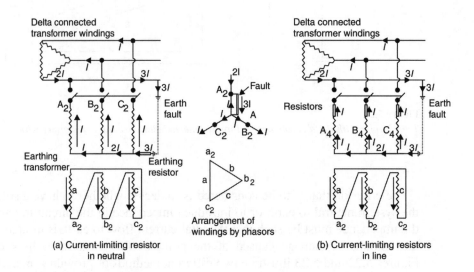

Figure 5.23
Three-phase, star/delta neutral earthing transformer

The value of these earthing resistors may be calculated simply by applying Ohm's law:

$$I = \frac{V\sqrt{3}}{Z_N}$$

$$Z_N = \frac{V\sqrt{3}}{I}$$

Frequently used devices for earthing high-voltage neutrals are the liquid neutral earthing resistor (LNER), and are capable of withstanding currents up to 1500 A for about 30 s. This type of resistor required regular maintenance like topping up of the electrolyte and maintaining its strength and fluidity in extreme winters by employing heaters, etc. To overcome this shortcoming, solid metal resistors may be used; of course this flexibility does come with additional costs.

As an alternative to earthing resistors and the associated problems, many substations use an arc suppressor. This device called a Petersen coil is a reactor connected between the neutral of the supply transformer and earth, and its reactance is tuned to match the capacitance of the power system it is protecting. Figure 5.24 shows a typical oscillogram of voltage recovery after arc suppression.

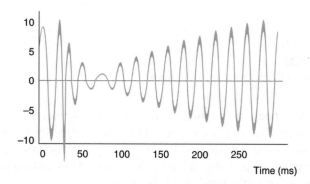

Figure 5.24
Recovery voltage after initial arc extinction

In the event of a sustained fault condition, the arc suppression coil allows the power system to be operated in a faulted condition until the fault can be located and removed. Nowadays solid-state tap switching control devices are employed in conjunction with arc suppression coils to achieve more efficient compensation.

In some instances when the high-voltage winding is common with the low-voltage winding for economic advantages, such a type of transformer is known as an autotransformer. These are most exclusively used to interconnect very high-voltage systems. Three-phase autotransformers are invariably star/star-connected and their use requires that the systems that they interconnect are able to share a common earthing arrangement (Figure 5.25).

Transformers with tertiary windings

It is most common to come across transformers with a third winding called tertiary winding in addition to primary and secondary windings. This is adopted in large size transformers in power stations and large switching substations. The most common reason for the additional third set of windings to a three-phase transformer is for the provision of a delta-connected tertiary winding to provide path for the third harmonics. Other reasons are:

- To limit the fault level on the LV system by subdividing the infeed – i.e. double secondary transformers
- The interconnection of several power systems operating at different supply voltages
- The regulation of system voltage and of reactive power by means of a synchronous capacitor connected to the terminals of one winding.

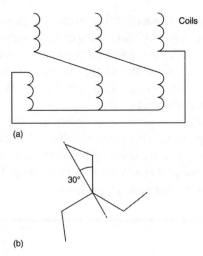

Figure 5.25
Interconnected star winding arrangement

As has been mentioned earlier in this chapter, it is desirable that a three-phase transformer has one set of three-phase windings delta-connected, thus providing a path for third harmonic currents. The presence of delta-connected windings also allows current to circulate around the delta in the event of an unbalance in the loading between the phases, so that this unbalance is reduced.

If the neutral of the supply and the star-connected winding are both earthed, then although the transformer output wave form will be undistorted, the circulating third-order harmonic currents flowing in the neutral can cause interference with telecommunications circuits and other electronic equipment. Further it can cause unacceptable heating in any neutral earthing resistors, so this provides an added reason for the use of a delta-connected tertiary winding.

If the neutral of the star-connected winding is unearthed then, without the use of a delta tertiary, this neutral point can oscillate above and below earth at a voltage equal in magnitude to the third-order harmonic component. Because the use of a delta tertiary prevents this, it is sometimes referred to as a stabilization winding.

The load currents in the primary phases corresponding to a single-phase load on the secondary of a star/star transformer with a delta tertiary are typically as shown in Figure 5.26 assuming a one-to-one turns ratio for all windings.

This leads to an ampere-turns rating of the tertiary approximately equal to one-third that of the primary and secondary windings and provides a rule-of-thumb method for rating the tertiary in the absence of any more specific rating basis.

Double secondary transformers

This is another type of multiwinding transformers where it is required to split the number of supplies from a HV feeder to economize on the quantity of HV switchgear and at the same time limit the fault level of the feeds to the LV switchgear.

In designing the double secondary transformer, it is necessary that both LV windings are disposed symmetrically with respect to the HV winding so that both have identical impedances to the HV. The two possible arrangements to achieve this, is as shown in Figure 5.27 (both instances there is a crossover between the two LV windings half way up the limb).

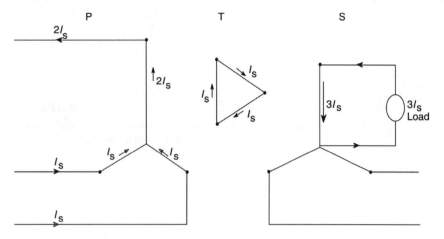

Figure 5.26
Single-phase load to neutral

The leakage reactance LV_1 to LV_2 in the first case would be high as the LV windings of Figure 5.27(a) are loosely coupled due to the fact the inner LV upper half crosses to the outer upper half and the inner lower half crosses to the outer lower half.

The leakage reactance LV_1 to LV_2 in the second case would be low as the LV windings of Figure 5.27(b) are closely coupled due to the fact the upper inner crosses to lower outer and upper outer crosses to lower inner. Thus, the equivalent circuits for these two arrangements are as in Figure 5.28 wherein the transformer is represented by a three-terminal network and typical values of impedance (leakage reactance) are marked.

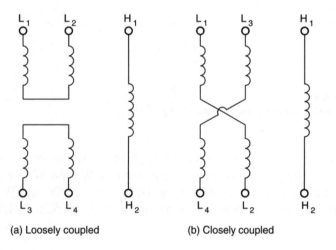

(a) Loosely coupled (b) Closely coupled

Figure 5.27
Transformers with two secondary windings

General case of three-winding transformer

The basic voltage for each winding and any combination of loading is the no-load voltage obtained from its turns ratio. In considering the case of two output windings W_2 and W_3, and one input winding W_1 as shown in Figure 5.29, there are three possibilities of loading conditions:

W$_2$ only loaded
W$_3$ only loaded
W$_2$ and W$_3$ both loaded

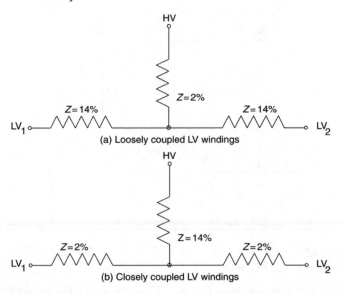

Figure 5.28
Equivalent circuits for loosely coupled and closely coupled double secondary transformers

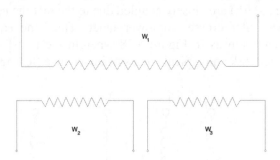

Figure 5.29
Diagram of a three-winding transformer

Now based on each of these three loading conditions two separate values are calculated viz. the regulation of each output winding W_2 and W_3 for a constant voltage applied to winding W_1, and the voltage regulation between W_2 and W_3 relative to each other. From this data an equivalent circuit, as shown in Figure 5.30, is derived as follows:

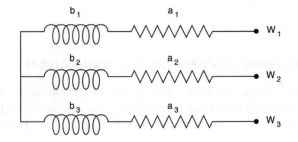

Figure 5.30
Equivalent circuit of a three-winding transformer

Let a_{12} and b_{12} be the percentage of resistance and reactance voltage referred to the basic kVA respectively and obtained from test, short circuiting either winding W_1 or W_2 and supplying the other with winding W_3 on open circuit,

Similarly, a_{23} and b_{23} by test to windings W_2 and W_3 with W_1 on open circuit, similarly, a_{31} and b_{31} by test to windings W_3 and W_1 with W_2 on open circuit,

$$d = (a_{12} + a_{23} + a_{31}), \text{ and}$$
$$g = (b_{12} + b_{23} + b_{31})$$

Then calculating for a_1, a_2, a_3, b_1, b_2 and b_3 we get:

Arm W_1: $a_1 = d/2 - a_{23}$ and $b_1 = g/2 - b_{23}$
Arm W_2: $a_2 = d/2 - a_{31}$ and $b_2 = g/2 - b_{31}$
Arm W_3: $a_3 = d/2 - a_{12}$ and $b_3 = g/2 - b_{12}$

The regulation with respect to the terminals of any pair of windings is the algebraic sum of the regulations of the corresponding two arms of the equivalent circuit.

Following is an example calculation of voltage regulation of a three-winding transformer given:

W_1 is a 66 000 V primary winding

W_2 is a 33 000 V output winding loaded at 2000 kVA and having a power factor $\cos \Phi_2 = 0.8$ lagging.

W_3 is an 11 000 V output winding loaded at 1000 kVA and having a power factor $\cos \Phi_3 = 0.6$ lagging.

In addition, the following test data is also available, related to a basic loading of 1000 kVA.

$a_{12} = 0.26$ and $b_{12} = 3.12$
$a_{23} = 0.33$ and $b_{23} = 5.59$
$a_{31} = 0.32$ and $b_{31} = 5.08$

Hence

$d = 0.91$ and $g = 9.79$

Then for

W_1: $a_1 = 0.125$ and $b_1 = +3.305$
W_2: $a_2 = 0.135$ and $b_2 = -0.185$
W_3: $a_3 = 0.195$ and $b_3 = +5.775$

The effective full-load kVA input to winding W_1 is:

1. With only the output winding W_2 loaded, 2000 kVA at a power factor of 0.8 lagging
2. With only the output winding W_3 loaded, 1000 kVA at a power factor of 0.6 lagging
3. With both the output windings W_2 and W_3 loaded, 2980 kVA at a power factor of 0.74 lagging.

Applying expressions (5.9) or (5.10) separately to each arm of the equivalent circuit, the individual regulations have, in

W_1 under condition (i) where $n_1 = 2.0$, the value of 4.23%
W_1 under condition (ii) where $n_1 = 5.0$, the value of 2.72%
W_1 under condition (iii) where $n_1 = 2.98$, the value of 7.15%
W_2 where $n_2 = 2.0$, the value of -0.02%
W_3 where $n_3 = 5.0$, the value of 5.53%

Summarizing these calculations, therefore the total transformer voltage regulation has:

1. With output winding W_2 fully loaded and W_3 unloaded
   At the terminals of winding W_2, the value of $4.23 - 0.02 = 4.21\%$
   At the terminals of winding W_3, the value of $4.23 + 0 = 4.23\%$
2. With output winding W_2 unloaded and W_3 fully loaded
   At the terminals of winding W_2, the value of $2.72 + 0 = 2.72\%$
   At the terminals of winding W_3, the value of $2.72 + 5.53 = 4.25\%$
3. With both output windings W_2 and W_3 fully loaded
   At the terminals of winding W_2, the value of $7.15 - 0.02 = 7.13\%$
   At the terminals of winding W_3, the value of $7.15 + 5.53 = 8.68\%$.

Terminal markings of transformers

To ensure proper design and installation, individual phase windings are given descriptive letters and the same letter in combination with suffix numbers is then used for all windings of one phase. The HV windings are given upper case letters and the LV windings are given lower case letters for corresponding phases.

Thus for single-phase transformers:

A   : for the HV winding
3A   : for the third winding (if any)
a   : for the LV winding

For two-phase windings on a common core or separate cores in a common tank:

AB   : for the HV windings
ab   : for the LV winding

For three-phase transformers:

ABC: for the HV winding
abc : for the LV winding

Figure 5.31 shows the standard markings of a single-phase transformer.

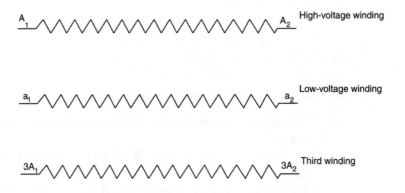

Figure 5.31
Terminal marking of a single-phase transformer having a third winding

Figure 5.32 shows the relative mar kings of a two winding transformers.

In addition to the letters marking the terminals, suffix numbers are given in sequence, relative to the direction of the induced emf, on all the tapping points and the ends of the winding. In the case of three-phase star-connected transformers, A_1 would be connected to the star's neutral point and the A_2 would be the line terminal. Figure 5.33 illustrates typical terminal markings of tappings.

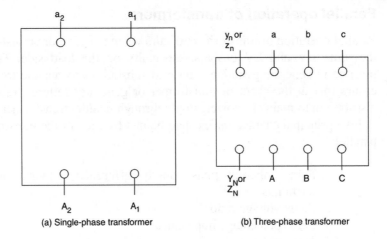

Figure 5.32
Relative position of terminals of two-winding transformers

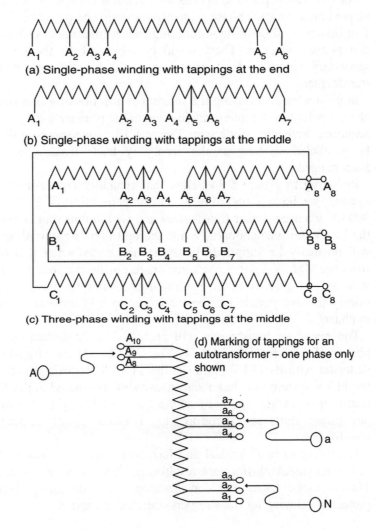

Figure 5.33
Marking of tappings on phase windings

Parallel operation of transformers

Parallel operation of transformers occurs when two or more transformers are connected to the same HV and LV busbars, especially on the load side. This is a most common practice in generating stations and grid substations to operate transformers in parallel to ensure that in the event of transformer or generator failures, redundancy is built in the distribution to maintain power supply through healthy transformers.

Five principal characteristics shall be met by each of the transformers to be operated in parallel:

1. Same inherent phase angle difference between primary and secondary terminals
2. Same voltage ratio
3. Same percentage impedance
4. Same polarity
5. Same phase sequence.

For two transformers to operate in parallel it is imperative that in addition to all of the above conditions, the phase sequence must also be the same. Figure 5.17 earlier showed four instances of delta/star-connected transformer under different conditions of polarity and phase sequence. There could be cases where the phase sequence required in secondary side of one transformer could be different with respect to the other transformer.

In such a case, reversing the internal connections on one side of the transformer can change polarities but interchanging two of the primary supply leads can reverse the phase sequence. Referring to the same Figure 5.17, transformers in diagrams in (a) and (d) can be paralleled so long as the secondary leads from a_1 and c_1 to the busbars are interchanged.

Referring to Figure 5.34 below that typically represents a power station auxiliary system, we have a 660 MW generator generating at 23.5 kV that is stepped up to 400 kV via a generator transformer and with a unit transformer providing a supply to the 11 kV unit switchboard. While the unit is being started up, the 11 kV switchboard will normally be supplied via the station transformer that takes its supply directly from the 400 kV system. At some stage during the loading of the generator, supplies will be changed from the station transformer to the unit transformer, which will briefly involve paralleling these sources so it is imperative that the supplies must be in phase.

The generator transformer will probably be connected star/delta with the 23.5 kV phasor at 1 o'clock, i.e. YNd1. The 23.5/11 kV unit transformer will be connected delta/star with its 11 kV phasor at the 11 o'clock position, i.e. Dyn11. This means that the 11 kV system now has a zero-phase shift compared to the 400 kV system. Thus, the station transformer must now connect the 400 kV system to the 11 kV system without any phase shift, and therefore this is achieved by utilizing a star/star-connected transformer.

Paralleling of two identical transformers reduces the combined impedances to half that of the individual transformers, resulting in the increase of the fault level of the LV busbar. This is to be taken into consideration while designing distribution equipment and protective relaying for transformers operated in parallel.

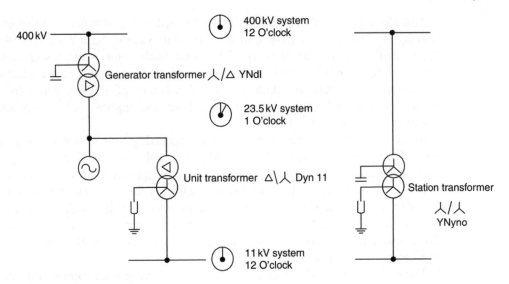

Figure 5.34
Power station auxiliary system

5.3   Transformer construction and installation

Transformer construction

Transformer construction comprises of windings and magnetic core kept in a closed tank and only windings brought out for external termination purpose. The closed tank construction is essential for safety as well as protection of the internal parts of the transformer. Further it is necessary to keep the high-voltage windings insulated from the other phases and the external surface. The most common construction is oil-filled transformer where the windings are immersed in a mineral oil having certain basic electrical characteristics like dielectric break down voltage, moisture content, etc. The oil also acts as the cooling medium which by natural circulation through the transformer radiators transfers the heat to the ambient. This construction is termed as ONAN (oil natural, air natural). Different types can be adopted to achieve higher cooling by external forced fans (ONAF) and by pumping oil using external devices (OFAF).

The other type of transformer is dry type and as the name implies this construction does not employ oil but use impregnated insulation to act as the insulating medium. The construction is air cooled by natural ventilation or with external fans.

Layout and installation practice

In designing the layout and installations of transformers, the following need to be considered: the location of the transformer whether it is to be installed outdoors or indoors, ventilation, fire hazard, transformer earthing and noise, etc.

Typically, most transformers that are oil filled present a potential fire hazard, especially if the transformer contains BS 148 oil. Therefore, it is absolutely essential to consider this factor when deciding whether a transformer is going to be installed outdoors or indoors. Invariably, most oil-filled transformers are therefore usually installed outdoors, or when essential to have the installation indoors, then adequate ventilation and physical isolation is required. This is to ensure, that in the unfortunate event of the transformer oil igniting for whatever reason, the damage caused shall be restricted to the transformer alone and its immediate ancillary equipment, and shall not interfere with any other unit assemblies in its vicinity.

There is a general misconception that all oil-filled transformers are fire hazards. This is usually not the case for most mineral oil-filled systems, which are less of a fire hazard than most. In these transformers, the closed flash point is not lower than 140 °C and hence it shall not be possible to accumulate sufficient vapor in an enclosed space to be ignited upon exposure to a flame or any other source of ignition, below this temperature. In any case, in order to sustain a flame, these oils require a wick to produce sufficient vapor to enable it to burn freely.

In the past usually there had to be a fault occurring prior to a fire that caused the rapid loss of oil through a rupture in the tank. Invariably, there will exist a high temperature, and as the oil is at this point exposed to the atmosphere, ignition will occur and then the transformer insulation will now serve as a wick for the oil and sustain combustion. On the other hand, if on the occurrence of a fault that causes a slow leak or drip of the oil onto a heated surface, if the temperature is high enough to ignite the oil/vapor, this fire could be fed continuously by the slow drip or leak. In many instances, this may go unnoticed until it escalates to a serious level.

There are many methods of avoiding or at the very least minimizing the risk of such hazards. Conventionally, it has been a general practice in many substations that employ oil-filled transformers and switchgear, to provide surfaces of chippings and a drainage sump to transport away any oil spillage that could potentially fuel a fire. This method however is not foolproof. It has been found that over a period, these chippings collect dust and grime and this grime would provide the wick for sustained combustion. Alternatively, one could explore the possibility of providing a firewater sprinkler system, which could be automatically triggered on the event of a fire around the transformer. Note that water displaces oil and any old oil spillage would be washed up from the sump. Therefore, it is imperative that this large quantity of excess oil and water be removed as quickly as possible to deluge water treatment centers before allowing the water to enter into the storm water drains, devoid of oil. Provision should be made to separate oil from water and for containment before this separation.

Additional precautions may also be considered if the above solutions are not feasible, such as separation or segregation. Separation involves locating the transformer well away from all other equipment, but this may not be convenient, as there may be space constraints. On the other hand, segregation provides for firewalls to be built around transformers such that fires, if any, would be contained within the walls. This firewall or barrier must be suitably reinforced to be capable of withstanding any explosions from the transformer.

In order to install transformers within buildings, it is a general practice that one uses dry-type resin-encapsulated units rather than using liquid-filled transformers.

In selecting the dielectric type the following must be adhered to:

- The dielectric must be non-toxic, biodegradable and must not present a hazard to the environment.
- The dielectric must have a fire point above 300 °C to be classified as a fire resistant fluid.
- The dielectric must not contribute to or increase the spread of an external fire nor must the products of combustion be toxic.
- Normal operation, electrical discharges or severe arcing within the transformer must not generate fumes or other products that are toxic or corrosive.

This does not mean that liquid-filled transformers cannot be used within buildings. They would meet all of the above criteria and in addition are cheaper and smaller than cast-resin or other dry type units. However there must be provision for a total spillage of the dielectric with suitable sumps and/or bunded catchment areas, such that in the event that spillage occurs the building drains would not be flooded with the dielectric liquids. If the transformers are installed on higher levels, then suitable precautions must be taken to prevent leakages on the lower floors.

On the other hand the building must be made totally weatherproof and care taken to ensure that there would be no deluges due to pipe leaks on the dry type of transformers after installation. Needless to say, every installation should have proper ventilation.

Noise in transformers

By definition, noise is described as an unpleasant or unwanted sound. It is inevitable that a transformer in operation emits noise. Transformer noise is continuous and it usually falls into the mid ranges of human audio spectrum. Invariably the noise that tends to irritate humans emanates from distribution transformers that are required to be located near homes and offices. Methods to attenuate this type of noise comes from employing various noise absorbing barriers such as existing walls, buildings, natural geographic topology (pits, artificial or natural), or even installing the transformer downwind from dwellings.

5.4   Transformer protection

Transformers are important in any distribution system and it is essential that they provide continuous service and are protected against faults that may keep out of service for long durations. Before proceeding with the protection methods, following paragraphs highlight the properties of transformers which shall be kept in mind while designing the protective system.

Transformer magnetizing characteristics

For efficiency reasons transformers are generally operated near to the 'knee-point' of the magnetic characteristic. Any increase above the rated terminal voltage tends to cause core saturation and therefore demands an excessive increase in magnetization current.

When a transformer is energized, it follows the classic magnetization curve as shown in Figure 5.35.

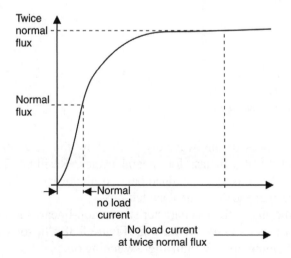

Figure 5.35
Transformer magnetizing characteristics

In-rush current

Under normal steady-state conditions, the magnetizing current, *I*, required to produce the necessary flux is relatively small, usually less than 1% of full load current (Figure 5.36):

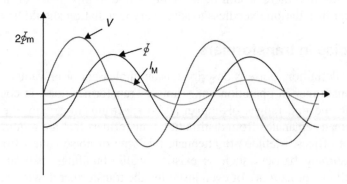

Figure 5.36
Steady-state conditions

However, if the transformer is energized at a voltage zero then the flux demand during the first half voltage cycle can be as high as twice the normal maximum flux. This causes an excessive unidirectional current to flow, referred to as the magnetizing in-rush current as shown in Figure 5.37.

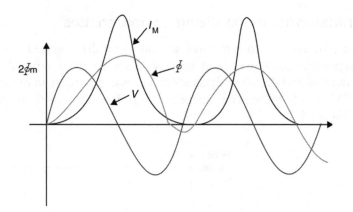

Figure 5.37
Illustration of magnetizing in-rush current

An analysis of this waveform will show that it contains a high proportion of second harmonic and will last for several cycles. Residual flux can increase the current still further, the peak value attained being of the order of 2.8 times the normal value if there is 80% permanence present at switch-on.

As the magnetizing characteristic is non-linear, the envelope of this transient in-rush current is not strictly exponential (Figure 5.38). In some cases, it has been observed to be still changing up to 30 min after switching on.

It is therefore important to be aware of these transient phenomena when considering differential protection of transformers, which will be discussed later.

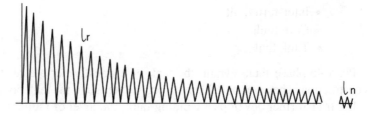

Figure 5.38
Typical transient current-rush when switching in a transformer at instant when $E = 0$

Mismatch of current transformers

Current transformers are provided on the HV and LV sides of a power transformer for protection purposes.

If we consider a nominal 132/11 kV 10 MVA transformer, the HV and LV full load currents would be as follows (Figure 5.39).

A ratio of 50/1 A would most likely be chosen for the HV current transformers, as it is not possible to obtain fractions of a turn. A ratio of 500/1 could be achieved comfortably for the LV current transformers.

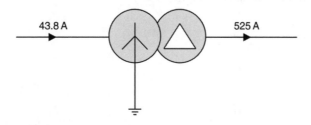

43.8 A   525 A

Figure 5.39
Nominal 132/11 kV 10 MVA transformer

We therefore have a mismatch of current transformer ratios with respect to the full load currents on the secondary side of the CTs (these are 0.876 A and 1.05 A). Incidentally 600/1 on the secondary side would be more suitable.

Furthermore, it is more than likely that the HV CTs and the LV CTs will be supplied by different manufacturers. There is therefore no guarantee that the magnetization curves will be the same, which will further add to the mismatch. This mismatch shall be duly considered while selecting the settings for the relays protecting the transformers.

Types of faults

A clear knowledge is required to understand the different types of faults that can occur in a transformer and their causes. This will enable the designer to adopt the correct protective system depending upon the possibilities for such faults, cost of protection vis-à-vis breakdown cost, etc. The following paragraphs briefly explain the descriptions for the following types of faults that can occur in a power transformer:

- HV and LV bushing flashovers (external to the tank)
- HV winding earth fault
- LV winding earth fault

- Inter-turn fault
- Core fault
- Tank fault.

Phase-to-phase faults within the tank of a transformer are relatively rare by virtue of its construction. They are more likely to occur external to the tank on the HV and LV bushings.

If a transformer develops a winding fault, the level of fault current will be dictated by:

- Source impedance
- Method of neutral earthing
- Leakage reactance
- Position of fault in winding (i.e. fault voltage).

Earth faults

Effectively earthed neutral

This refers to the solid connection of neutral to the earth. The fault current in this case is controlled mainly by the transformer leakage reactance, which varies in a complex manner depending on the position of the fault in the winding.

The reactance decreases toward the neutral so that the current actually rises for faults toward the neutral end (Figure 5.40).

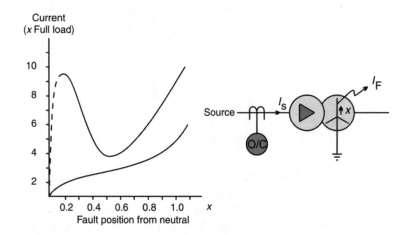

Figure 5.40
Relationship of fault current to position from neutral (earthed)

The input primary current is modified by the transformation ratio and is limited to 2 to 3 times the full load current of the transformer for fault positions over a major part of the star winding.

An overcurrent relay on the HV side will therefore not provide adequate protection for earth faults on the LV side.

Non-effectively (resistance or impedance) earthed neutral

For this application, the fault current varies linearly with the fault position, as the resistor is the dominant impedance, limiting the maximum fault current to approximately full load current (Figure 5.41).

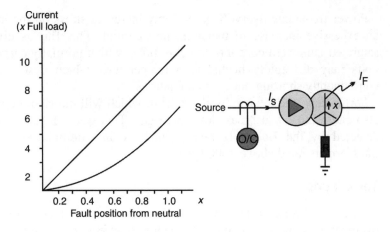

Figure 5.41
Relationship of fault current to positions from neutral (resistance earthed)

The input primary current is approximately 57% of the rated current making it impossible for the HV overcurrent relay to provide any protection for LV earth faults.

Restricted earth fault protection is therefore strongly recommended to cover winding earth faults and this will be covered in more detail in a later section.

Inter-turn faults

Insulation between turns can break down due to electro-magnetic/mechanical forces on the winding causing chafing or cracking. Ingress of moisture into the oil can also be a contributing factor.

Also an HV power transformer connected to an overhead line transmission system will be subjected to lightning surges sometimes several times rated system voltage. These steep-fronted surges will hit the end windings and may possibly puncture the insulation leading to a short-circuited turn (Figure 5.42). Very high currents flow in the shorted turn for a relatively small current flowing in the line.

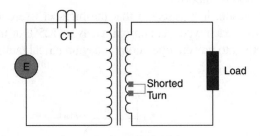

Figure 5.42
Inter-turn faults

Core faults

Heavy fault currents can cause the core laminations to move, chafe and possibly bridge causing eddy currents to flow, which can then generate serious overheating.

This additional core loss will not produce any noticeable change in the line currents and thus cannot be detected by any electrical protection system.

Power frequency overvoltage not only increases stress on the insulation but also gives an excessive increase in magnetization current. This flux is diverted from the highly saturated laminated core into the core bolts, which normally carry very little flux. These bolts may be rapidly heated to a temperature, which destroys their own insulation, consequently shorting out core laminations.

Fortunately, the intense localized heat, which will damage the winding insulation, will also cause the oil to break down into gas. This gas will rise to the conservator and be detected by the Buchholz relay, which is an external device provided in oil-filled transformers rated above 1000 kVA.

Tank faults

Loss of oil through a leak in the tank can cause a reduction of insulation and possibly overheating on normal load due to the loss of effective cooling.

Oil sludge can also block cooling ducts and pipes, contributing to overheating, as can the loss of forced cooling pumps and fans generally fitted to the larger transformer.

5.5   Relays for protection

HV overcurrent

The most common protection employed in a transformer is protection against withdrawal of currents from the source in excess of rated/design values. Also it is quite common that a ground fault or inter-phase faults in a transformer winding could be the cause for flow of fault currents many times the rated design current. The most common protection against such drawing of excess currents is by using an IDMTL (inverse definite minimum time) overcurrent and earth fault relay on the HV side of a transformer. The operating time varies inversely with respect to the current values and the operating time can be set to the desired values in these relays. The inherent time delay of the IDMTL element provides back-up for the LV side.

High-set instantaneous overcurrent is also recommended on the primary side mainly to give high-speed clearance to HV bushing flashovers. Care must be taken, however, to ensure that these elements do not pick-up and trip for faults on the LV side as discrimination is important.

For this reason, it is essential that the high-set overcurrent element should be of the low-transient over-reach type, set approximately to 125% of the maximum through-fault current of the transformer to prevent operation for asymmetrical faults on the secondary side (Figure 5.43).

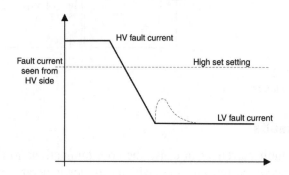

Figure 5.43
Fault current as seen from the HV side

This relay therefore looks into, but not through the transformer, protecting part of the winding, so behaving like unit protection by virtue of its setting.

When grading IDMTL overcurrent relays across a delta–star transformer it is necessary to establish the grading margin between the operating time of the star side relay at the phase-to-phase fault level and the operating time of the delta side relay at the three-phase fault level.

This is due to the fact that, under a star side phase-to-phase fault condition, which represents a fault level of 86% of the three-phase fault level, one phase of the delta side transformer will carry a current equivalent to the three-phase fault level.

Differential protection

Differential protection, as its name implies – compares currents entering and leaving the protected zone and operates when the differential current between these currents exceed a predetermined level.

The type of differential scheme normally applied to a transformer is called the current balance or circulating current scheme as shown in Figure 5.44.

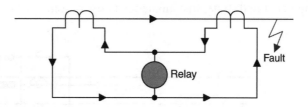

Figure 5.44
Differential protection using current balance scheme (external fault conditions)

The CTs are connected in series and the secondary current circulates between them. The relay is connected across the mid-point where the voltage is theoretically nil, therefore no current passes through the relay, hence no operation for faults outside the protected zone.

Under internal fault conditions (i.e. faults in area within the two sets of CTs) the relay operates because there is an unbalance detected by the relay within its protective zone as shown in Figure 5.45.(The CT currents tend to flow in the same direction through the relay making it to operate.)

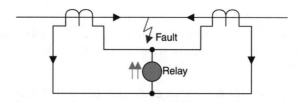

Figure 5.45
Differential protection and internal fault conditions

This protection is called UNIT protection, as it only operates for faults on the unit it is protecting, which is situated between the CTs. The relay therefore can be instantaneous in operation, as it does not have to coordinate with any other relay on the network.

This type of protection system can be readily applied to autotransformers (Figures 5.46 and 5.47) as follows.

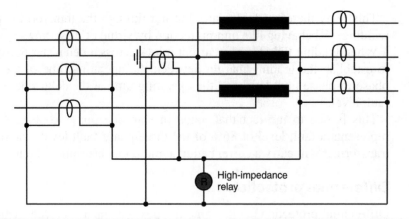

Figure 5.46
Restricted earth protection applied to autotransformers

All current transformer ratios remain the same and the relays are of the high-impedance (voltage operated) type, instantaneous in operation.

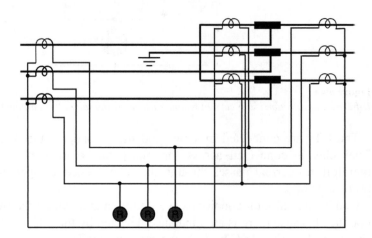

Figure 5.47
Autotransformer – phase and earth fault scheme

However it is not just enough to select CT ratio alone for adopting differential protection in a two-winding transformer. As explained earlier, there are a number of factors that need consideration:

1. Transformer vector group (i.e. phase shift between HV and LV)
2. Mismatch of HV and LV CT's characteristics
3. Varying currents due to on-load tap changer (OLTC)
4. Magnetizing in-rush currents (from one side only)
5. The possibility of zero sequence current destabilizing the differential for an external earth fault.

Factor (a) can be overcome by connecting the HV and LV CTs in star/delta respectively (or vice versa) opposite to the vector group connections of the primary windings, so counteracting the effect of the phase shift through the transformer.

The delta connection of CTs provides a path for circulating zero sequence current, thereby stabilizing the protection for an external earth fault as required by factor (e).

It is then necessary to bias the differential relay to overcome the current unbalances caused by (b) mismatch of CTs and (c) OLTC. And finally, as the magnetizing current in-rush is predominantly second harmonic, filters are utilized to stabilize the protection for this condition (d).

Most transformer differential relays have bias and slope setting of 20, 30 and 40% as shown. For OLTC-operated transformers, the operating range of the OLTC dictates the desired setting, which is responsible for the biggest current unbalance under healthy conditions.

E.g. If the OLTC range is +15 to –5% = 20% then a minimum of 20% bias setting is selected. The CT mismatch should also be taken into account.

Typical connections for a delta–star transformer would be as shown in Figure 5.48. Note that the CTs on primary are connected in star while those in secondary are connected on delta to offset the phase shift in the two windings.

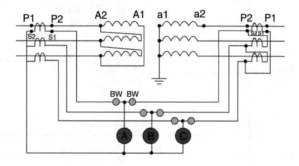

Figure 5.48
Typical connections for a delta–star transformer

Under load or through-fault conditions, the CT secondary currents circulate, passing through the bias windings to stabilize the relay, whilst only small out-of-balance spill currents will flow through the operate coil, not enough to cause operation. In fact the higher the circulating current, the higher will be the spill current required to trip the relay, as can be seen from the following characteristic in Figures 5.49 and 5.50.

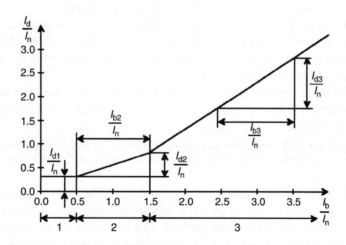

Figure 5.49
Operating current of differential relay

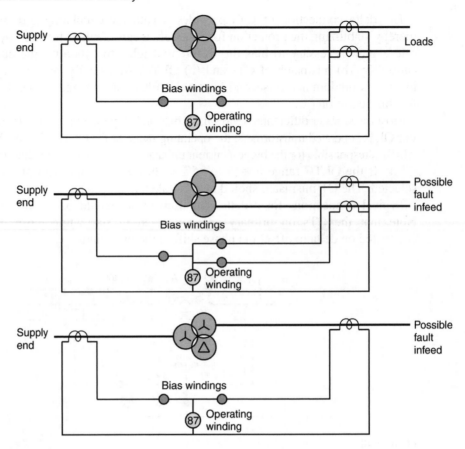

Figure 5.50
Biased differential configurations

Restricted earth fault

As demonstrated earlier in these notes, a simple overcurrent and earth fault relay will not provide adequate protection for winding earth faults.

Even with a biased differential relay installed, biasing can cause the relay to become ineffective for certain earth faults within the winding. This is especially so if the transformer is resistance or impedance earthed, where the current available on an internal fault is disproportionately low.

In these circumstances, it is often necessary to add some form of separate earth fault protection. The degree of earth fault protection is very much improved by the application of unit differential or restricted earth fault systems as shown in Figure 5.51.

On the HV side, the residual current of the three-line CTs is balanced against the output current of the CT in the neutral conductor, making it stable for all faults outside the zone.

For the LV side, earth faults occurring on the delta winding may also result in a level of fault current of less than full load, especially for a mid-winding fault which will only have half the line voltage applied. The HV overcurrent relays will therefore not provide adequate protection. A relay connected to monitor residual current will provide earth fault protection since the delta winding cannot supply zero-sequence current to the system.

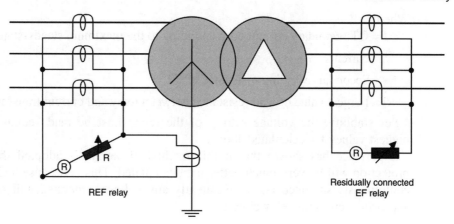

Figure 5.51
A restricted earth fault system

The relay used is an instantaneous high-impedance type, the theory of which is shown as follows:

Determination of stability

The stability of a current balance scheme using a high-impedance relay depends upon the relay voltage setting being greater than the maximum voltage which can appear across the relay for a given through-fault condition (Figure 5.52).

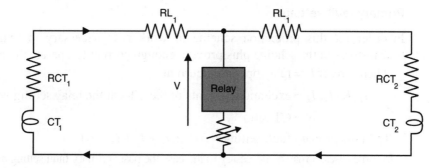

Figure 5.52
Basic circuit of high-impedance current balance scheme

This maximum voltage can be determined by means of a simple calculation, which makes the following assumptions:

- One current transformer is fully saturated, making its excitation impedance negligible.
- The resistance of the secondary winding of the saturated CT together with lead resistance constitute the only burden in parallel with the relay.
- The remaining CTs maintain their ratio.

Referring to Figure 5.52, the maximum voltage is given

$$V = I(R_{ct} + R1)$$

Where

I = CT secondary current corresponding to the maximum steady-state through-fault current

R_{ct} = Secondary winding resistance of CT

$R1$ = Largest value of lead resistance between relay and current transformer.

For stability, the voltage setting of the relay must be made equal to or exceed the highest value of V calculated above.

Experience has shown that if this method of setting is adopted the stability of the protection will be very much better than calculated. This is because a CT is normally not continuously saturated and consequently any voltage generated will reduce the voltage appearing across the relay circuit.

Method of establishing the value of stabilizing resistor

To give the required voltage setting the high-impedance relay operating level is adjusted by means of an external series resistor as follows:

Let v = operating voltage of relay element

Let i = operating current of relay equipment

and V = maximum voltage as defined under 'determination of stability' above.

It is sometimes the practice to limit the value of series resistor to say 1000 Ω, and to increase the operating current of the relay by means of a shunt connected resistor, in order to obtain larger values of relay operating voltage.

From the equation above $V = I(R_{ct} + R1)$

Therefore $R1 = (V/I) - R_{ct}$

Primary fault setting

In order for this protection scheme to work it is necessary to magnetize all current transformers in the scheme plus provide enough current to operate the relay.

Therefore, if I_r = relay operating current

I_1, I_2, I_3, I_4 = excitation currents of the CT's at the relay setting voltage

N = CT ratio

Then the primary fault setting = $N(I_r + I_1 + I_2 + I_3 + I_4)$

In some cases, it may be necessary to increase the basic primary fault setting as calculated above.

If the required increase is small, the relay setting voltage may be increased (if variable settings are available on the relay) which will have the effect of demanding higher magnetization currents from the CT's I_1, I_2, etc.

Alternatively, or when the required increase is large, connecting a resistor in parallel with the relay will increase the value of I_r.

Current transformer requirements

Class X CTs are preferably required for this type of protection; however experience has shown that most protection type CTs are suitable for use with high-impedance relays, provided the following basic requirements are met:

- The CTs should have identical turns ratio. Where turns error is unavoidable, it may be necessary to increase the fault setting to cater for this.
- To ensure positive operation, the relay should receive a voltage of twice its setting. The knee-point voltage of the CTs should be at least twice the relay

setting voltage (knee-point = 50% increase in mag. Current gives 10% increase in output voltage).
- CTs should be of the low-reactance type.

Protection against excessively high voltages

As the relay presents a very high impedance to the CTs, the latter are required to develop an extremely high voltage. In order to contain this within acceptable limits, a voltage dependant resistor (VDR), or metrosil, is normally mounted across the relay to prevent external flashovers (Figure 5.53), especially in polluted environments.

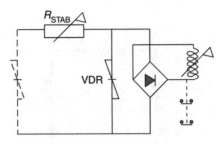

Figure 5.53
Protection against excessively high voltages

Example

Using Figure 5.54, calculate the setting of the stabilizing resistor for the REF protection. The relay is 1 A rated with 10–40% tappings, burden 1.0 VA.

$$\text{Secondary fault current} = 9000 \times \frac{1}{3000}$$
$$= 30 \text{ A}$$

Relay operating current

Choose 10% tap on the relay (10%) = 0.1 A

Relay operating voltage

$$\frac{VA}{I} = \frac{\text{burden}}{\text{current}} = \frac{1.0}{0.1} = 10 \text{ V}$$

Stabilizing voltage

$$V = I\left(R_{ct} + R1\right) = 30(3 + 1) = 120 \text{ V}$$

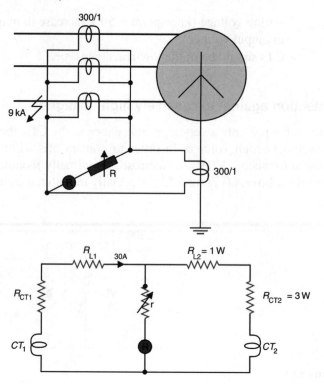

Figure 5.54
Example for calculation of setting of stabilizing resistor

Voltage across stabilizing resistor

$$\text{Stabilizing voltage} - \text{relay voltage} = 120 - 10 = 110 \text{ V}$$

$$\text{Stabilizing resistance} = \frac{\text{Voltage across resistor}}{\text{Current through resistor}}$$

$$= \frac{110}{0.1}$$

$$= 1100 \ \Omega$$

Current transformer must therefore have a minimum knee-point voltage of $2 \times 120 = 240$ V to ensure positive operation of protection for an internal fault.

6

Grounding, earthing and transient overvoltage protection

6.1   Grounding devices

Solid grounding

The neutral of a power transformer is grounded solidly with a copper conductor (see Figure 6.1).

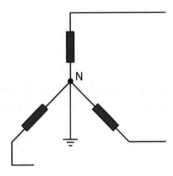

Figure 6.1
Solid grounding of power transformer

Advantages

- Neutral held effectively at ground potential
- Phase-to-ground faults of same magnitude as phase-to-phase faults so no need for special sensitive relays
- Cost of current-limiting device is eliminated
- Size and cost of transformers are reduced by grading insulation toward neutral point N.

Disadvantages

- As most system faults are phase-to-ground, severe shocks are more considerable than with resistance grounding
- Third harmonics tend to circulate between neutrals.

Resistance grounding

A resistor is connected between the transformer neutral and earth (see Figure 6.2).

- Mainly used from 6.6 to 33 kV
- Value is such as to limit an earth fault current to between 1 and 2 times full load rating of the transformer, typically 400 A.

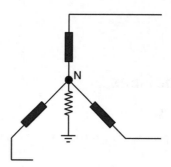

Figure 6.2
Resistance grounding

Advantages

Limits electrical and mechanical stresses on system when an earth fault occurs, but at the same time current is sufficient to operate normal protection equipment.

Disadvantages

Full line-to-line insulation required between phase and earth.

Reactance grounding

A reactor is connected between the transformer neutral and earth (see Figure 6.3).

- Values of reactance are approximately the same as used for resistance grounding.
- To achieve the same value as the resistor, the design of the reactor is smaller and thus cheaper.

However, reactance grounding can lead to severe overvoltages during arcing earth faults (see IEEE standard 142 and SABS 0200). This form of grounding is not recommended.

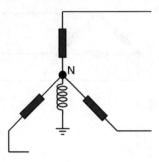

Figure 6.3
Reactance grounding

Arc suppression coil (Petersen coil)

A tuneable reactor is connected in the transformer neutral to earth (see Figure 6.4).

- Value of reactance is chosen such that reactance current neutralizes capacitance current. The current at the fault point is therefore theoretically nil and unable to maintain the arc, hence its name.
- Virtually fully insulated system, so current available to operate protective equipment is so small as to be negligible. To offset this, the faulty section can be left in service indefinitely without damage to the system as most faults are earth faults of a transient nature, the initial arc at the fault point is extinguished and does not restrike.
- Sensitive wattmetrical relays are used to detect permanent earth faults.

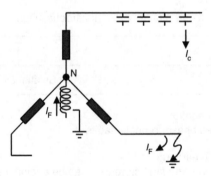

Figure 6.4
Arc suppression coil (Petersen coil)

Earthing via neutral electromagnetic coupler with resistor

This type of earthing provides an earth point for a delta system and combines the virtues of resistance and reactance grounding in limiting earth fault current to safe relayable values (see Figure 6.5).

The current should be resistive in nature to comply with IEEE standard 142 and SABS 0200. Comparisons are made in Figures 6.6 and 6.7.

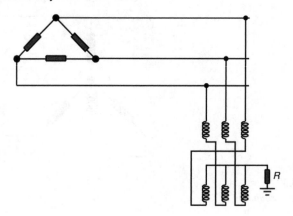

Figure 6.5
Earthing via neutral earthing compensator

Problems
Phase faults
High fault currents
Only limited by inherent impedance of power supply
Earth faults
Solid earthing means high earth fault currents
Only limited by inherent zero sequence impedance of power system
Consequences
Heavy currents damage equipment extensively – danger of fire hazard
This leads to long outage times – lost production, lost revenue
Heavy currents in earth bonding give rise to high touch potentials – dangerous to human life
Large fault currents are more hazardous in igniting gases – explosion hazard

Solutions
Phase segregation
Eliminates phase-to-phase faults
Resistance earthing
Means low earth fault currents – can be engineered to limit to any chosen value
Benefits
Fault damage now minimal – reduces fire hazard
Lower outage times – less lost production, less lost revenue
Touch potentials kept within safe limits – protects human life
Low fault currents reduce possibility of igniting gases – minimizes explosion hazard
No magnetic or thermal stresses imposed on plant during fault
Transient overvoltages limited – prevents stressing of insulation and breaker restrikes

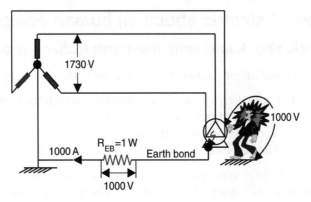

Figure 6.6
Touch potentials – solid grounding

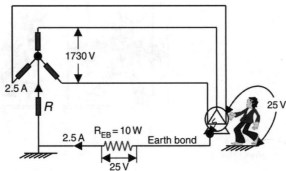

Figure 6.7
Touch potentials – resistive grounding

Evaluation of Relative Merits of Effective and Resistive Earthing (See IEEE standard 142 and SABS 0200)		
1 **Subject**	**2** **Effective Earthing**	**3** **Resistive Earthing**
Rated voltage of system components, particularly power cables and metal oxide surge arresters	Need not exceed $0.8U_m$	Must be at least $1.0U_m$ for 100 s
Earth fault current magnitude	Approximately equal to three-phase fault current (typically 2–10 kA)	Reduced earth fault current magnitude (typically 300–900A)
Degree of damage as a result of an earth fault	High degree of damage at fault point and possible damage to feeder equipment	Lesser degree of damage at fault point and usually no damage to feeder equipment
Step and touch potentials during earth fault	High step and touch potentials	Reduced step and touch potentials
Inductive interference on and possible damage to control and other lower voltage circuits	High probability	Lower probability
Relaying of fault conditions	Satisfactory	Satisfactory
Cost	Lower initial cost but higher long-term equipment repair cost	Higher initial cost but lower long-term equipment repair cost usually making resistive earthing more cost-effective

6.2   Effect of electric shock on human beings

Electric shock and sensitive earth leakage protection

There are four major factors, which determine the seriousness of an electric shock:

1. Path taken by the electric current through the body
2. Amount of current
3. Time the current is flowing
4. The bodies electrical resistance.

The most dangerous and most common path is through the heart (see Figure 6.8).

Persons are not normally accidentally electrocuted between phases or phase to neutral, almost all accidents are phase to earth.

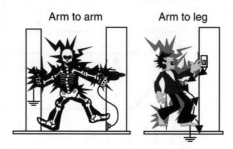

Figure 6.8
Dangerous current flows

Figure 6.9 shows the four stages of the effect of a current flow through the body:

1. Perception – tingling – about 1 mA
2. Let-go threshold level – about 10 mA
3. Non-let-go threshold level – 16 mA
4. Constriction of the therasic muscles – death by asphyxiation and ventricular fibrillation – about 70–100 mA.

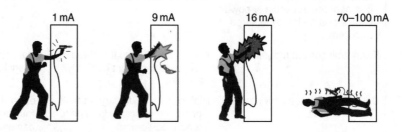

Figure 6.9
Effects of current flow through the body

Figure 6.10 shows the normal electrocardiogram – one pulse beat – at 80 bpm = 750 ms.

- QRS phase – normal pumping action
- T phase – refractory or rest phase – about 150 ms.

Death could occur if within this very short period of 150 ms a current flow was at the fibrillation level.

Figure 6.11 shows the resistance of the human body – hand to hand or hand to foot.

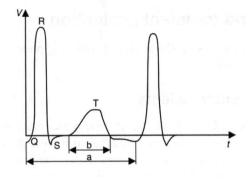

Figure 6.10
Electrocardiagram

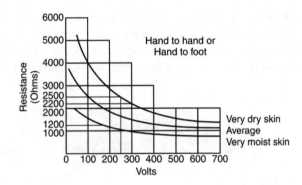

Figure 6.11
Resistance of human body

Consider an example of a man working, possibly perspiring, he touches a conductor at 300 V (525 V phase to earth).

300 V divided by 1000 Ω = 300 mA!

It is important to remember that it is current that kills, not voltage.

Sensitive earth leakage protection

Figure 6.12 illustrates the operation of the core balance leakage device.

This shows a single-phase system but would be similar for three-phase, 3w or three-phase, 4w systems.

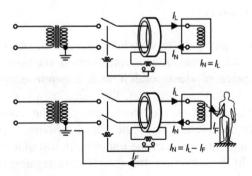

Figure 6.12
Principles of core balance protection

6.3   Surge and transient protection

This section focuses on the basics of lightning phenomenon, switching surges and typical mitigation techniques.

Earth current transients

In Figure 6.13 looking at the comparison between atmospheric transients and earth current transients, we are reminded that, if it were possible, it would do us a lot of good to be in the open air, above earth, rather than to be in the earth itself. We would much rather take our chances with a varying electrostatic field and the results that occur in an insulating medium, the air, rather than take our chances with what is taking place in a conducting medium, namely, the earth. In days gone by, we were able to suspend our telephone and data lines underbuilt below a power line. We were even able to hang our inter-plant lines in open air and the risk was much less from the direct strikes or the effects of the electrostatic field than it is when we go into the earth. We might have induced transients come to us as lines are close to each other in the construction process of underbuilding and overbuilding, but again, the end-result effects were considerably less.

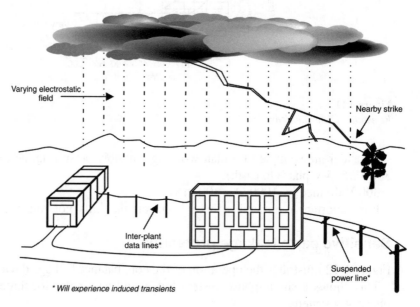

Figure 6.13
Atmospheric transients

Consider Figure 6.14 now, where placing all of our lines in the earth, we tend to no longer use the old-fashioned construction methods. The older methods consisted of bearing a piece of steel conduit in a concrete encasement and then placing wires in the center. In that construction, the metallic wires were protected, at least in part, by the shielding effect of the steel conduit. The conduit, in turn, was protected from deterioration in contact with the soil by being encased with concrete. With today's modern technology, most of the under earth installations are performed with PVC conduit with very little protection for the electromagnetic properties of the noise transients running in the earth. In Figure 6.15, we can see the impact of earth current transients running back and forth and perhaps intercepting the conductors running in the plastic conduit buried between the buildings. If there were direct lightning strikes to the earth or if a power line with a lightning arrestor was discharging its energy into the earth, then the PVC conduit with wires might be intercepted and the metallic wires would carry these transient charges in both directions.

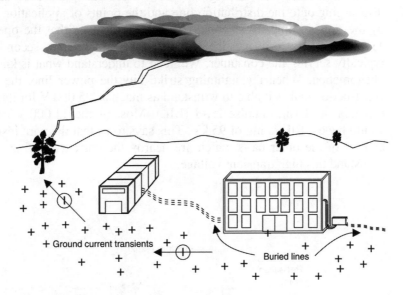

Figure 6.14
Earth current transients

The one exception to this, of course, is glass fiber installations which do not, of themselves, conduct any of these earth transients when buried in the earth directly.

Transient impact

In Figure 6.15 we see the impact of a transient phenomenon making for high field strength and dielectric stress as well as a very high rate of change of voltage with regard to time, producing a high current component. Just a reminder to us: transient phenomena can be very dangerous.

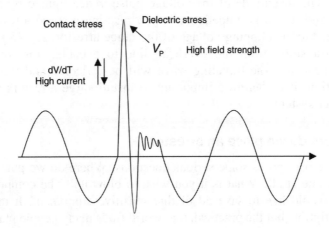

Figure 6.15
Impact of a transient

Lightning phenomenon

Let us begin by talking about what happens when a lightning strike hits an overhead distribution line. Here in Figure 6.16 we see the picture of the thunderstorm cloud discharging onto the distribution line and the points of application of a lightning arrestor by the power company at points 1 and 2. We notice that the operating voltage here is 11 000 V on the primary line and the transformer has a secondary voltage of 400 V typically serving the consumer. We need to understand what is known as traveling wave phenomenon. When the lightning strike hits the power line, the power line's inherent construction makes it able to withstand as much as 95 000 V for its insulation system. We call this the basic impulse level (BIL). Most of the 11 000 V construction equipment would have a BIL rating of 95 kV. This says to us that the wire insulation, the cross arms and all of the other parts which are nearby the current-carrying conductors are able to withstand this high transient voltage.

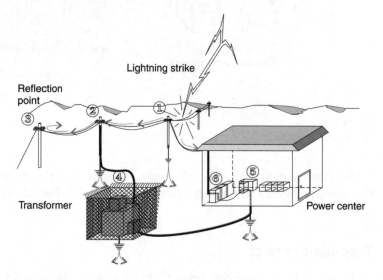

Figure 6.16
Lightning strike to an overhead distribution line

When an overhead line is struck by lightning, an overvoltage surge travels down the line. The magnitude of the voltage pulse is determined by the current of the lightning volt and the surge impedance. Upon reacting a surge arrestor the overvoltage will be limited to the clipping voltage of the range arrestor, say 45 kV for an 11 kV system. At the transition from the overhead line to a cable, transformer or at the end of the overhead line the traveling wave will be reflected and the voltage doubled by super position. It is therefore important to install surge arrestors at all transition points in a power system.

Where do we place an arrestor?

Figure 6.17 poses some serious questions. Where do we put a lightning arrestor? Do we need one at all? What is it you wish to blow up? The computer? The UPS? Or perhaps you would like to save all of that sensitive equipment. It may be a humorous type of description, but the practical, real world finds many people placing lightning arrestors and discharge devices inside of occupied spaces where the equipment and the personnel represent a heavy financial burden as well as a safety factor. Most safety standards will advise that discharge devices do not need to be located where there are either personnel or equipment to be protected. The location of a discharge device, such as a lightning arrestor, is to be at the large service entrance earth, where the electric utility makes its service connection to the premises. Here, at this point, this discharge device which has large levels of current, then has a sufficiently large earth plane into which to discharge that current without a damaging effect on sensitive equipment. Typical lightning arrestor ratings call for 65 000 A of discharge capacity for distribution class arrestors, 100 000 A of discharge capacity for station class arrestors, and even at the 600 V and below level, 40 000 A of capability at the minimum. So we notice that these questions are not silly, but they do point out that we need to locate our lightning arrestor product as close to the service entrance as possible.

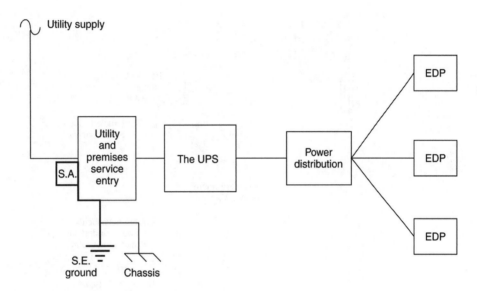

Figure 6.17
Positioning of a lightning arrestor

Arrestor and supplementary protection

From the Federal Information Processing Standards we have a picture of the combined placement of lightning arrestor products and surge protection devices called transient surge protectors. Notice that the writers of this document were very accurate in the location of the arrestor product and to make it as close to the power source as possible. In addition, they give us a picture of the use of the older style arresting products, which required capacitors to affect wavefront modification. Let us explain that. Wavefront modification means that the voltage rise is so fast that if something does not mitigate that rise, the wiring may be bridged by the extremely high voltage in the surge.

Downstream from the arrestor location, over a certain amount of distance, preferably greater than 10–15 m (30–50 ft), if possible, should be the second level of protection shown in Figure 6.18 as a transient surge protector. This device, indicated as combination suppresser and filter package made up of a variety of different types of components which will now protect against the residual energy that is flowing in the circuit. The structure that we see here is one in which the various components installed in the system, starting at the service entrance, then to a sub-panel and then, finally, to discrete individual protectors, will now attenuate more and more of the surge energy until it is completely dissipated.

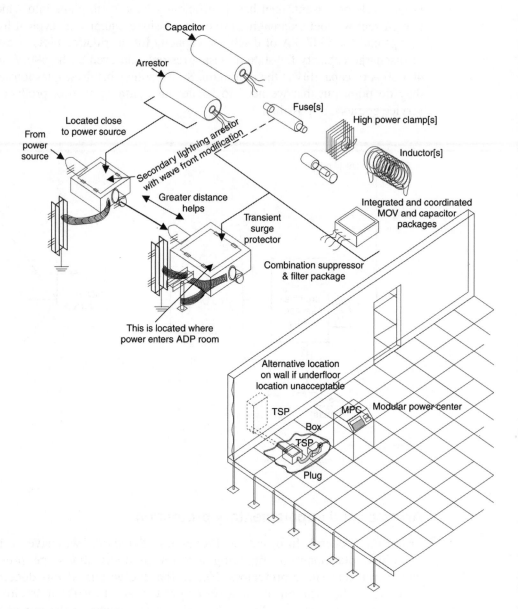

Figure 6.18
Supplementary surge protection elements

Transient clipping

In Figure 6.19, we see the onrushing surge coming into a transient voltage surge suppression device (TVSS), now becoming better known as surge protection device (SPD), and having its upper portion limited or chopped before it comes to the load. The surge protection element actually is taking off or limiting that high-transient phenomenon and sending it back into the earth path where it will do very little harm.

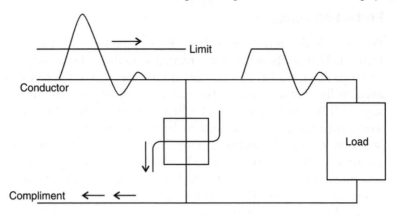

Figure 6.19
Onrushing surge

Protection locations

Figure 6.20 consists of a series of diagrams demonstrating the overall electric utility supply and internal wiring. We see the recommendation that is consistent with what we have been talking about before. The black boxes marked on our drawing as SPDs first appear connected at the service entrance equipment inside the building where it receives power from the service transformer. Next we see an SPD device at a panel board or sub-panel assembly. Finally, we may find a lower voltage style device as a discrete device either plugged in at an outlet or perhaps approaching the mounting of this device within a particular piece of sensitive equipment itself.

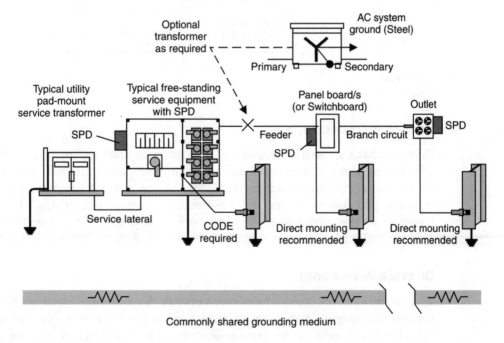

Figure 6.20
Typical location of power distribution TVSS/SPD

Protection zones

We call this the zoned protection approach and we see these various zones outlined in Figure 6.21 with the appropriate reduction in the order of magnitude of the surge current as we go down further and further into the zones, into the facility itself. Notice that in the uncontrolled environment outside of our building we would consider the amplitude of say, 1000 A. As we move into the first level of controlled environment, called zone 1, we would get a reduction by a factor of 10 to possibly 100 A of surge capability. As we move into a more specific location, zone 2, perhaps a computer room or where the various sensitive hardware exist, we find another reduction by a factor of 10. Finally, in the equipment itself, we may find another reduction by a factor of 10 of the effect of all of this surge being basically 1 A at the device itself.

The idea of the zone protection approach is to utilize the inductive capacity of the facility, namely the wiring, to help attenuate the surge current magnitude as we go further and further away from the service entrance to the facility.

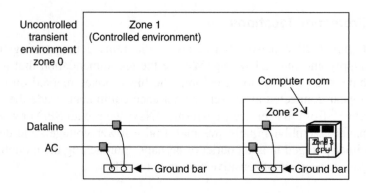

	Zone 0	Zone 1	Zone 2	Zone 3
Order of magnitude of surge current	x 1000 A	x 100 A	x 10 A	x 1 A

Figure 6.21
Zoned protection approach

Service entrance zone

Here we have a detailed picture of the entrance into the building where both the telecommunications, data communications and the power all enter from the outside to the first protected zone (see Figure 6.22). Notice that SPD is basically stripping any transient phenomena on any of these metallic wires, referencing all of this to the common service entrance earth even as it is attached to the metallic water piping system.

Discrete device zone

Here as we address the discrete level between the first level of controlled zone 1 and perhaps the plug-in device taking it into the zone 2 location, we can see SPDs are available that handle the telecommunications, data and different types of physical plug connections for each, including both the RJ type of telephone plug as well as coaxial wiring (see Figure 6.23).

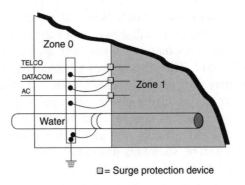

Figure 6.22
Zone 0 protection

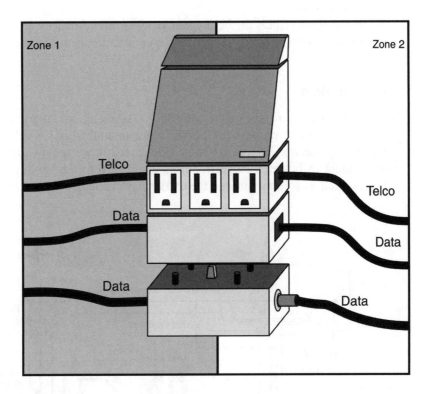

Figure 6.23
Zones 1 and 2

This is a common design error where two points of entry and therefore two earthing points are established for the AC power and telecommunications circuits. The use of SPDs at each point is highly beneficial in controlling the line-to-line and line-to-earth surge conditions at each point of entry, but the arrangement cannot perform this task between points of entry. This is of paramount importance since the victim equipment is connected between the two points. Hence, a common-mode surge current will be driven through the victim equipment between the two circuits despite the presence of the much needed SPDs. The minimal result of the above is corruption of the data and maximally, there may be fire and shock hazard involved at the equipment.

No matter what kind of SPD is used in the above arrangement nor how many and what kind of additional individual, dedicated earthing wires, etc. are used, the stated problem will remain much as discussed above. Wires all possess self-inductance and because of $-e = L$ di/dt conditions, cannot equalize potential across themselves under normal impulse/surge conditions. Such wires may self-resonate in quarter-waves and odd-multiples thereof, and this is also harmful. This also applies to metal pipes, steel beams, etc. Earthing to these nearby items may be needed to avoid lightning side-flash, however.

Adding more earths to the above (such as at the victim equipment) is not effective since these merely act as taps into the voltage–current divider that the earth represents due to its ohms/unit value at any given frequency. This value gives up with increased frequency. Therefore, such taps just create more loops for the surge currents to flow and ring back-and-forth in. These connections cannot be implemented in any case when the equipment is located on floors above earth. Therefore, schemes based upon such connections would fall in any case due to the need to place equipment anywhere in a building and on any floor, etc. and the 'long' wires necessitated to be used to reach 'earth' floors away.

While the shown SPDs are highly beneficial (from normal-mode surge), some additional improvements must be made to cope with common-mode problems (such as lightning!).

Site overview

We are fortunate to have from consulting associate, Mr Warren Lewis, an excellent overview of how to prepare the power, telecommunications, earthing and surge current routing for protection. In our first view (Figure 6.24), we see the typical victim equipment being caught in between surge impulses coming from the service entrance of electrical power as well as coming through the entrance of the telecommunications product.

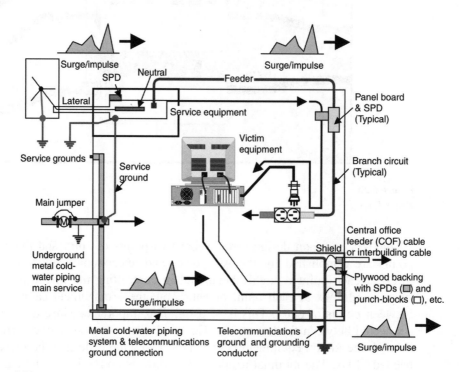

Figure 6.24
Overview of a typical site: Power and telecommunications earthing and surge current routing through victim equipment

Otherwise unchanged from the previous drawing, Figure 6.25 shows the recommended usage and location of a combination data and AC SPD unit employed for protection of the equipment connected between the AC and telecommunications systems. This practice is easy to implement on a case-by-case basis, but it is not a complete solution since the beneficial effects of the arrangement are not complete. In other words, there is always a residual surge current to deal with due to Kirchkoff's current law being operative at the junctions within the AC and data SPD unit. No SPD vendor can eliminate these phenomena.

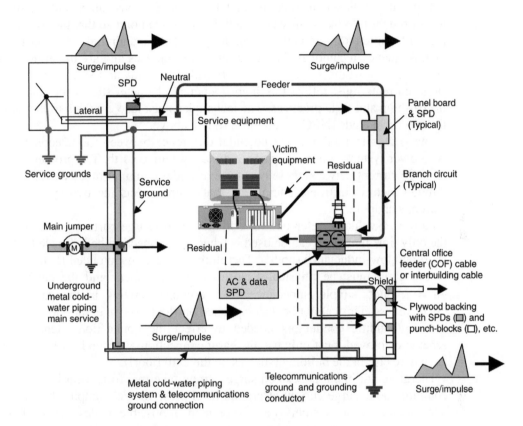

Figure 6.25
Improved protection of victim equipment connected between AC power and telecommunications circuits (localized SPD protection)

The SPD located at the service entry is beneficial in that it diverts some of the current that would otherwise enter the victim equipment's susceptibility path. The same holds true for the SPD at the telecommunications entry area. Both are important to have.

The shown solution is still flawed since the surge current is still allowed to flow largely within the facility and on the AC and telecommunications wiring. This means lots of stray coupling into other circuits along the way, and in some cases, localized arcing between items that are conductive and at different potentials. The locally installed SPD at the AC and telecommunications ports for the shown 'Victim Equipment' cannot have any practical effect on these unwanted phenomena. Therefore, something further must be done to improve matters to the level consistent with the demands of a modern, commercial establishment utilizing electronic computer/data processing and telecommunications systems that are interconnected.

Specialized 'earth' earthing at the local SPD point is not generally required of beneficial, in fact, adding another 'earth' into the circuit at this point might be an additional surge current injection point! Local connection to local building steel, etc. are always a good practice, but not a requirement.

Improved protection

The improvement in protection for the victim equipment is shown in Figure 6.25 where the protected devices are located at the incoming power, a sub-panel board and the victim equipment itself.

Typically for new construction, the telecommunications cables and the AC power need to be brought into the facility at essentially the same point so that the two may be literally locked-together from an earthing/bonding and SPD standpoint. This must be done using high-frequency techniques. Surge currents are impulses and are therefore high-frequency currents. As a result, wiring practices intended to motivate the unwanted effects of such surge currents must reflect this high-frequency characteristic in the implementations of the earthing and SPD practices. The techniques must also conform to the requirements of the Electrical Code (NEC) as do those shown above.

The goal is to minimize the potential difference between the telecommunications and AC power systems, not necessarily of either system to earth. The problem is the exchange of surge current between these two systems under common-mode current conditions, and through victim electronic equipment, that interconnects the two systems from a surge current standpoint.

Earth is a valuable reference for common-mode surge current (such as lightning), but the only time that earth can be used is when the connection being made to it is physically short and low-inductance (both!). Therefore, connections intended to divert current in/out of the earth must be physically close to the equipment and the earth point itself or $-e = L$ di/dt, simple impedance, and resonance effects in the path will defeat its effectiveness. This cannot be accomplished on floors above earth due to the length of the earthing/bonding conductors needed to effect the connection. Hence, the protection scheme employed must only use the earth connection at ground level – it is not effective anywhere else in the building (such as on the 25th floor!).

Therefore, all initial attacks on surge current take two forms: earth ground diversion of common-mode surge current at ground level plus voltage clamping between the AC and telecommunications systems (must be done in all possible modes). This is best done at the common point of entry for both systems to the facility. Then, secondary and tertiary attachments on the common-mode surge current to a place whenever the two systems (ac power and telecommunications) can be physically brought together and SPD interconnected in low-inductance fashion. Normal-mode protection on the AC system and the telecommunications system is generally effective wherever it is done in a path via the simple addition of a suitable SPD, since 'earth ground' is not a part of this process.

Recommended practice – joint entrance

Here in our Figure 6.26 of this sequence we find the recommended practice for installation of power and telecommunications entrances, typically, for new construction.

Two major changes are made in this figure to greatly reduce the surge current problem between the AC power and telecommunications circuit. The first is to disconnect the COAX cable from its landings on the plywood board and to reroute its end (including the gas-tube SPD assembly on it) into a metal pull-box. The cable's shields and the SPD are well-bonded to the metal box, which in turn is connected to a newly installed, driven earth ground installed just below the pull-box (this keeps wires short).

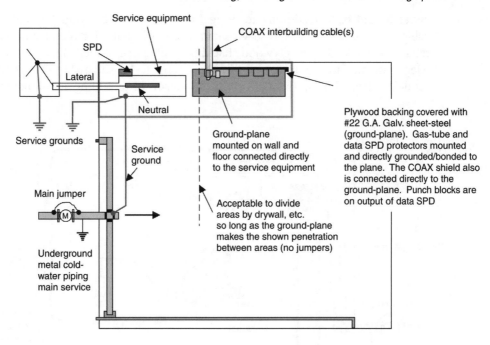

Figure 6.26
Recommended practice for installation of AC power and telecommunication lines on a facility (with SPD protection)

The second change is to reroute the COAX cable from the pull-box via ferrous metal conduit, to the service equipment area where it is landed to a high-quality SPD (i.e. a 'data protector') that is in turn earthed/bonded to the service equipment via a wall and floor mounted earth-plane. This is a #22 gage, galvanized steel arrangement that is secured to the wall, run down to the floor, bent at 90° and run to the service equipment, where it is lineally bonded to that equipment (weld, braze, solder or several screws, in that order or desirability). All COAX cable conduit is earthed/bonded to this earth-plane on both sides.

Next, the COAX cable is returned to the telecommunications area via another ferrous metal conduit as shown. Once the return is made the various pairs are landed to whatever punch-block terminals that they would have originally gone to.

The ferrous metal conduit used above is most effective if it is rigid conduit, then intermediate conduit, and finally, 'thinwall' (EMT) conduit. Do not use flexible conduits of any kind. Aluminum is acceptable, but steel is best. Also, do not make any effort to 'insulate' these conduits along the runs. Just install them in 'normal' fashion. In fact, the conduit is meant to connect multiple points of 'earth', the one at the service lateral (the power supply earth) and the other 'signal earth' at the communication entry point, into a single equipotential reference plane. When joining circuits or terminating them, use high-quality fittings and connectors. The best type is threaded, then circumferential compression ring types, and lastly, the set-screw types.

The service and telecommunications systems are SPD 'surge-locked together' in this arrangement.

Retrofitting recommendations

Here we find a recommendation for the retrofitting of a telecommunications and power system for improving the lightning and surge immunity (see Figure 6.27). You will notice that the heart of this is to correct the facility where a separate telecommunications entrance had been made and to bring ferrous conduit carrying the telecommunications cable main runs over to the service entrance earth-plane for the electric power and then return it back to its plywood backing and punch-block location at the separate telecommunications entrance. While this seems to be somewhat of a redundant method for doing these things, it works very well in terms of providing additional protection.

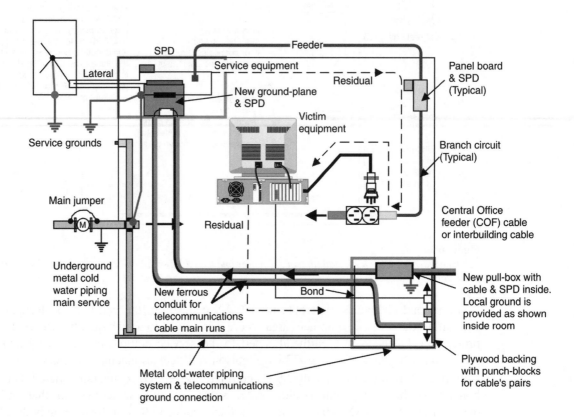

Figure 6.27
Retrofitting a telecommunication and AC power system for improved lightning and surge immunity (Recommended)

Still further improvement in surge protection is obtained once additional protection for the victim equipment is added at the local point where the equipment is installed. This is generally accomplished via the use of a combination AC and data SPD as shown and as previously discussed.

In the shown arrangement above, there is only 'residual' surge current to be controlled by the locally applied AC and data SPD as opposed to its being required to handle principal surge currents. This means that the locally applied protection becomes very much more effective since it is now acting as a second (and sometimes third) stage of surge current diversion and voltage clamping point. Less surge current is input to the local protection by the above arrangement so less escapes its effects to finally reach the victim equipment. However, something will always reach the victim equipment since no protection scheme is 'perfect'. But, if what does reach it, stays well below the point of component damage, the equipment will be 'protected' in all but the most extreme cases.

This arrangement is a form of a 'cascade' protection scheme. The idea is to progressively divert surge currents at each level – nearest the point of entry, then downstream, and finally at the victim equipment itself. At each level of protection, destructive energy is taken out of the surge, its current and voltage level is reduced, and its waveshape is also rendered less likely to have fast transitions which are the principal cause of $-e = L \, di/dt$ problems in the current's path.

Note that the original 'Telephone Company' protector (the gas-tube SPD) that is installed on the end of the COAX cable is retained and is locally earthed/bonded to both the metal equipment conduit/raceway system and to a local earth earthing electrode system. This maintains the 'Bell System' requirements for the installation and also permits a 'first line of defense' to be provided at the point of entry for the COAX to the building. This cable-end SPD may do a lot of good and can relieve the SPD installed at the service from having to handle the highest levels of surge currents that might enter via the telecommunications cable, etc. However, this does not mean that the 'data' SPDs installed at the service do not have to be 'heavy duty', they do. Surges arriving on the AC service will trigger these SPDs, and so they must handle the large amounts of current that this implies.

Adding local protection

Hence, in summary, to the routing of the telecommunications cable in ferrous conduit, we will add, as we have seen before, the local power, sub-panel and discrete surge protection devices to improve the immunity of the product (Figure 6.28).

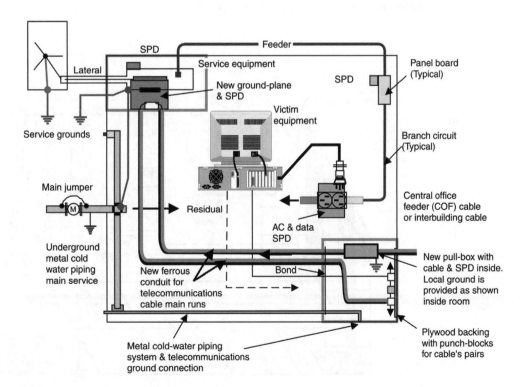

Figure 6.28
Adding local AC and data SPD protection for improved lightning and surge immunity (Recommended)

The earth-plane is now shown fully developed and covering both the wall(s) and the floor in the telecommunications equipment area. The whole area need not be covered by the earth-plane, just as shown/needed. Also, the AC power is now integrated with the same high-frequency earth system that the telecommunications system is using and all SPDs are commonly referenced to it as well.

The separately derived AC system needs to be solidly earthed as shown from one or more of the chosen AC system earthing electrodes (A, B or C). The preference is structural building steel, the made electrode system and then the metal cold-water piping, all in order of desirability. Best results are obtained when two or more of the shown electrodes are available to the earth-plane, although only one needs to be connected directly to the transformer for AC system earthing.

Cable trays/ladders must be electrically continuous so ends must be earthing/bonding jumpered together when two more sections meet or approach across a gap or other space. This is especially important to do when the tray/ladder penetrates walls, etc. Continuation of the cable into the site without tray/ladder is best done within ferrous metal conduit/raceway with it well earthed/bonded to the tray/ladder at the point of conversion.

Clamps, welds, brazing and direct soldering are all methods employed to connect items together instead of inductive means such as long earthing/bonding wires, etc. Such wires may be added to what is shown so as to allow the installer to 'see' the earthing.

Fully integrated protection

Reviewing the use of the earth-plane for a fully developed and fully integrated system, we find all of the components that we have been referring to now located and referenced to a common equipotential surface (Figure 6.29).

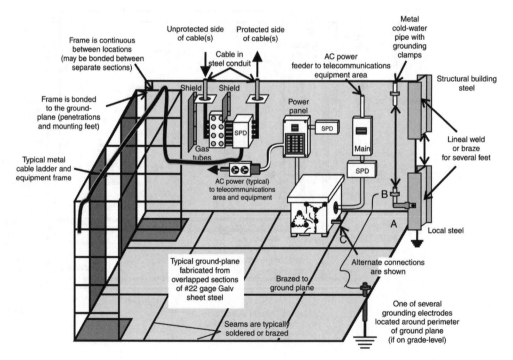

Figure 6.29
Fully integrated SPD protection and earthing for the telecommunication system

Circuit protective elements and transient phenomena

Fuses have been, until recently, the prime protective element in medium- and low-voltage circuits, with current ratings from as low as 60 mA–1000 A or more. These protective elements can carry up to 5000 times their rated current for period of <5 ms. These devices are manufactured with a wide range of characteristics that vary the speed of response (failure mode) with varying magnitudes of fault or overload currents vs time. Long-term failures can occur, caused by a continuous 20% overload, which will not become evident for hours or days. The separation of the wire element during the rupture period (clearing of the overload) can cause surges due to the arcing across the separating pieces of the element. Many fuses have the wire element surrounded in silica as a current suppressant, for 'current limiting' action, and the rapid melting of the silica as it surrounds the ruptured element leads to enormous transient voltages across the opening element. (The impedance of the wire rupture becomes nearly 'infinite' instantaneously!)

These phenomena cause stress in the associated fuses in adjoining phases of a three-phase system, and may lead to 'single phasing' in the system for indeterminate periods. The protected AC circuits tend to 'sag' just before the rupture by an amount depending on the fault current and source impedance, and the 'surge' after the complete rupture, particularly if the rupture time period is very short. The circuit will experience interference in the form of transient noise between the time of the current overload and the final clearing by the fuse element. The nominal ratings tend to show wide tolerances, as much as +/–20% is not uncommon, and thus this form of protection may not be accurate.

The general trend is to replace the fuse with a magnetic circuit breaker (MCB), capable of protecting all phases even when one phase only is effected. These devices exhibit only a resistive element in the circuit being protected, and offer additional protection in the form of earth leakage detection.

Earth leakage devices (single and three phase)

These devices operate by comparing the supply currents to the load return currents in a circuit. A difference in current, in excess of the mA rating of the device, results in the device operating to open all circuits, both phase(s) and neutral. A three-phase device can be obtained for three phase, three-wire operation or for three phase with neutral. When there is a leakage to earth, whether from any phase or neutral, the device will operate to protect the circuit, thus no connection between neutral and earth can exist downline of the unit. These units present a small series inductance on each line (phase and neutral) of the circuit.

Codes and guidelines

We find in Table 6.1 examples of various codes, standards and guidelines that are available for the understanding of standards of transient voltage surge suppression. What has been referred to by manufacturers in the United States as TVSS, now is more easily referred to throughout the international community as a SPD.

Organization	Code, Article or Standard No.	Scope
ANSI/IEEE	C62	Guides and standards on surge protection
	C62.41 – 1980	Guide for surge voltages in low voltage AC power circuits
	C62.1	IEEE standard for surge arrestors for AC power circuits
	C62.45 – 1987	Guide on surge testing for equipment connected to low voltage AC power circuits
	C62.41 – 1991	Recommended practice on surge voltages connected to low voltage AC power circuits (approved, not published)
IEEE	C74.199.6 – 1974	Monitoring of computer installations for power disturbances. International Business Machines Corp. (IBM)
UL	UL 1449	Transient Voltage Surge Suppressors (TVSS)
NEC	Article 250	Earthing
	Article 280	Surge arresters
	Article 645	Electronic data processing equipment
	Article 800	Communications circuits
NFPA	NFPA-75–1989	Protection of electronic data processing equipment
	NFPA-78 – 1989	Lightning protection code
	NFPA-20 – 1990	Centrifugal fire pumps
MIL-STD	MIL-STD-220A	50 Ω insertion loss test method. Earthing, bonding & shielding for electronic equipment
	MIL-STD-419A/B	and facilities
FIPS	FIPS PUB 94	Guideline for electrical power in ADP Installations (Chapter 7)

Table 6.1
Codes, recommended practices, standards and guidelines for transient voltage surge suppression (TVSS)/(SPD)

7

Power system analysis software

7.1   Introduction

Power system analysis software ranges from basic, commercially available, generic packages to large, complex programs developed for a specific customer. The latter serves the needs of a single comprehensive electrical system and may be integrated in real time to an electrical SCADA system with real input from the field.

The former programs are usually more user-friendly, require simulated input, and are used on a wide scale by consultants, industry and utilities.

A wide variety of programs are available today to perform all sorts of electrical analyses, as will be demonstrated in this chapter. Of all these, the programs that are undoubtedly used most frequently in power distribution design and analyses are the load flow and fault analyses studies.

A load flow study determines the voltage, current, power and reactive power in various points and branches of the system under simulated conditions of normal operation. Load flow studies are essential in optimizing existing networks, ensuring an economical and efficient distribution of loads, and plan future networks or additions to existing networks.

The currents that flow into different parts of a power system immediately after the occurrence of an electrical fault differ substantially from the current flowing in steady-state conditions. These currents determine the ratings of circuit breakers and other switchgear that are installed in the system, specifically the current flowing immediately after the fault and the current which the circuit breaker must interrupt. Fault calculations consist of predicting these currents for various types of faults at various locations in the system. The data obtained from fault calculations are also used to determine relay settings.

Network studies used to be extremely time-consuming in the pre-digital computer days, and often not very accurate. However, very powerful electrical analysis software in modern times has restricted human efforts to the input of equipment data and the interpretation of the study results. Network studies can now be done in a fraction of the time it used to take, with much more accurate results. This has greatly aided in more efficient power systems being designed, implemented and maintained.

The real-time information that can be obtained from a power system automation system is a great aid to, and parallel with, network studies. Firstly, the steady-state information continuously obtained, archived and trended can tremendously assist the results of load flow studies. Secondly, the information obtained from disturbance records, as well as sequence-of-events recordings, can be used to determine the correctness of fault calculations and stability studies, as well as to verify the effectiveness of the electrical protection.

7.2 Load flow

One of the earliest programs to be developed for power system analysis was the load flow (power flow) program. It was originally developed in the late 1950s. All complete packages in use today have load flow programs as an integral part, as this is one of the cornerstones of any electrical network analysis.

Although an electrical network is linear, load flow analysis is iterative due to the interactive influences of voltages, currents, frequency and reactance on one another. The generators are scheduled to deliver a specific active power to the system and usually the voltage magnitude of the generator terminals is fixed by automatic voltage regulation. Usually, one generator busbar only has its voltage magnitude specified, as losses in the system cannot be determined before the load flow solution. This bus also has its voltage angle defined to some arbitrary value, usually zero, which is only a reference point for the rest of the network. This busbar is known as the slack bus, or utility bus in some programs. The slack bus is a mathematical requirement for the program and has no exact equivalent in reality.

The total load plus the losses are not known in operating practice, and they fluctuate continuously. When a system is not in power balance, i.e., when the input power does not equal the load power plus losses, the imbalance modifies the rotational energy stored in the system. The system frequency thus rises if the input power is too large and falls if the input power is too little. Usually a generating station with one machine is given the task of keeping the frequency constant by varying the input power.

The algorithms originally developed had the advantages of simple programing and minimum storage but were slow to converge requiring many iterations. The introduction of ordered elimination of the network matrix, and programing techniques that reduce storage requirements, allowed much better algorithms to be used. The Newton–Raphson method gave convergence to the solution in only a few iterations.

Using Newtonian methods of specifying the problem, a Jacobian matrix containing the partial derivatives of the system at each node can be constructed. The solution by this method has quadratic convergence. This method was followed quite quickly by the fast decoupled Newton–Raphson method. This exploited the fact that under normal operating conditions, and providing that the network is predominately reactive, the voltage angles are not affected by reactive power flow and voltage magnitudes are not effected by real power flow. The fast decoupled method requires more iterations to converge but each iteration uses less computational effort than the Newton–Raphson method.

The results of a load flow study can be seen as a snapshot of the system once steady-state conditions have been reached with all loads staying at constant values. The study calculates the voltage on each bus, the voltage drop on each feeder, the current, power flow and losses in all branches, as well as total system losses. Branch loads can include load diversity considerations.

7.3 Sizing study

The sizing study selects the conductor size based on the minimum conductor material and cross-sectional area necessary to meet defined feeder-current-carrying capability, and associated voltage drop criteria. Transformers are sized based on their full load kVA rating.

Feeder sizing

Feeder sizes are based on the design load value from the demand load study. The demand load study calculates the total connected demand and design load in each branch of the power system. Some loads are defined as continuous loads and, as such, the design load value is larger than its demand load value. The NEC (USA) requires branch circuits that serve continuous loads to be rated so that not more than 80% of the feeder current rating is used. Selecting a design load value of 125% (1/80) of the demand load value meets the NEC standard. The feeder current rating is defined as the current in amperes that a conductor may carry continuously under the conditions of use without exceeding its temperature rating.

The sizing study bases its calculations on two separate criteria: the minimum conductor cross-sectional area to meet feeder rating values and a user-defined voltage drop value. If you consider parallel feeder combinations for a specified conductor type, the sizing study can select multiple feeders in parallel for that conductor. Cable sizes are usually specified in a cable library in order to be available to the sizing study. Derating factors are determined based on the temperature derating factor and duct bank design detail criterion.

The sizing study selects the cable that best meets defined current rating values and has the smallest cross-sectional area. Once the cable is selected, the voltage drop for the cable or cable pair is calculated. If the voltage drop criterion is exceeded, the sizing study selects the next larger cable size and begins the comparison of cross-sectional area, rated current and voltage drop.

The sizing study algorithm determines the feeder branch design load value in amperes, then determines the selected feeder design current rating. The feeder design rating is the product of the rated current, the temperature derating factor and the number of parallel cables. The design load value is the rated size of the load multiplied by specified demand factors and the long continuous load factor (or design factor).

Once the current rating conditions are met, the sizing study then checks the calculated voltage drop on the cable, based on the branch design load current and power factor, cable impedance, and length.

The voltage drop is calculated using the following formula:

$$\% \text{ Voltage drop} = \frac{\sqrt{3}\left[\left(I \times R \times \cos\Phi\right) + \left(I \times X \times \sin\Phi\right)\right]}{V_{LL}} \times 100$$

If the voltage drop exceeds the specified level, the sizing study selects the next largest feeder and restarts the sizing study in order to select a cable or combination of parallel cables which meets the rating criteria of minimum cross-sectional area and acceptable voltage drop.

Transformer sizing

Transformer sizing is normally based on either the calculated branch demand or design load value, depending on user selection. The sizing algorithm compares the demand or design load value to the transformer's full load size. The transformer's full load size is:

Full load size = Nominal kVA rating × Transformer capacity factor

Typical capacity factors are:

Transformer Cooling Characteristic	Capacity Factor
Dry type (DT)	1.00
Oil/air cooled (OA)	1.15
Oil/air/forced air (OAFA)	1.25

Transformer feeders

Feeders for the primary and secondary of transformers are based on a factor (usually 125%) of the transformer's full load rating and an allowable voltage drop criteria. This factor is usually defined in the demand load study set-up dialog box.

7.4   Fault analysis

A fault analysis program derives from the need to adequately rate switchgear and other busbar equipment for the maximum possible fault current that could flow through them. Fault analysis programs were also developed in the 1950s alongside load flow programs.

Initially only three-phase faults were considered and it was assumed that all busbars were operating at nominal voltage prior to the fault occurring. The load current flowing prior to the fault was also neglected.

By using the results of a load flow prior to performing the fault analysis, the load currents can be added to the fault currents allowing a more accurate determination of the total currents flowing in the system.

Unbalanced faults can be included by using symmetrical components, mathematically dividing fault currents in negative sequence, positive sequence and zero sequence currents.

When the fault levels in an industrial plant are calculated the contributions of motors and generators should also be taken into account. Their contribution to the fault current can change the final value significantly. Modern programs also take the DC effect into account.

7.5   Transient stability

After a disturbance, due usually to a network fault, the synchronous machine's electrical loading changes and the machines speed up (under very light loading conditions they can slow down). Each machine will react differently depending on its proximity to the fault, its initial loading and its time constants. This means that the angular positions of the rotors relative to each other change. If any angle exceeds a certain threshold (usually between 100° and 140°) the machine will no longer be able to maintain synchronism. This almost always results in its removal from service.

Early work on transient stability had concentrated on the reaction of one synchronous machine coupled to a very large system through a transmission line. The large system can be assumed to be infinite with respect to the single machine and hence can be modeled as a pure voltage source. The synchronous machine is modeled by the three phase windings of the stator plus windings on the rotor representing the field winding and the eddy current paths. These are resolved into two axes, one in line with the direct axis of the rotor and the other in line with the quadrature axis situated 90° (electrical) from the direct axis. The field winding is on the direct axis. Equations can be developed which determine the voltage in any winding depending on the current flows in all the other windings. A full set of differential equations can be produced, which allows the response of the machine to various electrical disturbances to be found. The variables must include rotor angle and rotor speed. The great disadvantage with this type of analysis is that the rotor position is constantly changing as it rotates. As most of the equations involve trigonometrical functions relating to stator and rotor windings, the matrices must be constantly re-evaluated.

In the most severe cases of network faults the results, once the DC transients decay, are balanced. Further, on removal of the fault, the network is considered to be balanced. There is thus much computational effort involved in obtaining detailed information for each of the three phases, which is of little value to the power system engineer. By contrast, this type of analysis is very important to machine designers. However, programs have been written for multi-machine systems using this method.

Initially, transient stability programs all ran in the time domain. A set of differential equations is developed to describe the dynamic behavior of the synchronous machines. All the machine equations are written in the direct and quadrature axes. The network is represented in the real and imaginary axes, similar to that used by the load flow and fault analysis programs.

Modern programs also look at the response of the system, not only to major disturbances but also to the build-up of oscillations due to small disturbances (such as asynchronous resonance) and poorly set control systems. In calculating the system response for these disturbances, very long time domain solutions are not suitable, so frequency domain models of the system were developed.

7.6   Fast transients

Transient stability program considers the dynamic response of the power system in the range of 1–10 Hz. There is however also a need to calculate the system's response to faster transients from switching ranges to the propagation of steep voltage and current waves traveling down transmission lines as a result of a lightning strike. To model these effects a very detailed model is required, e.g. all the stray capacitances and inductances need to be incorporated into the model.

Many of these programs are difficult to use and take a long time to solve. For this reason only short time periods are usually studied.

7.7   Reliability

Reliability of equipment is of constant concern to the operators of power systems. In the past, reliability was ensured by providing reserve equipment, either connected in parallel with other similar devices or which could be easily connected in the event of a failure.

However, providing reserve equipment to cater for all eventualities has become very costly as systems expand.

Reliability of a system is governed by the reliability of all the parts and their configuration. Much work has been done on the determination of the reliability of power systems but work is still being done to comprehensively model power system components and integrate them into system reliability models. Much of the early work was focused on generation facilities. The reasons for this was that, first, more information was available about the generation; second, the geographical size of the problem was smaller; and, third, the emphasis of power systems was placed in generation. With the onset of deregulation (in the USA), distribution and customer requirements are considered paramount.

At the generation and transmission levels, the loss of load expectation and frequency and duration evaluation are prime reliability indicators.

The usual method for evaluating reliability indicators at the distribution level, such as the average interruption ratio per customer per year, is an analytical approach based on a failure-mode assessment.

7.8   The second generation of programs

The advent of the PC provided a universal platform on which most users and programs could come together. This process was further assisted when window-based programs reduced the need for such a high level of computer literacy on the part of users.

PCs, Macintosh computers – which has a similar capability to a PC but which is less popular with engineers – and more powerful workstations, usually based on the Unix operating system, are used for software programs. Minicomputers and mainframe computers are also still in general use in universities and industry.

Hardware and software for power system operation and control required at utility control centers are usually sold as a total package. These are usually proprietary systems and the justification for a particular configuration requires input from many diverse groups within the utility.

7.9   Graphics

Two areas of improvement that stand out in this second wave of generally available programs are both associated with the graphical capabilities of computers. A good diagram can be more easily understood than many pages of text or tables.

The ability to produce graphical output of the results of an analysis has made the use of computers in all engineering fields, not just power system analysis, much easier. Tabulated results are never easy to interpret. They are also often given to a greater degree of accuracy than the input data warrants. A graph of the results, where appropriate, can make the results very easy to interpret and if there is also an ability to graph any variable with other variables, especially if three dimensions can be utilized, then new and significant information can be quickly assimilated.

New packages became available for business and engineering which were based on either the spreadsheet or the database principle. These also had the ability to produce graphical output. It was no longer essential to know a programing language to do even quite complex engineering analysis. The programing was usually inefficient and obtaining results was more laborious, e.g. each iteration had to be started by hand. But, as engineers had to use these packages for other work, they became very convenient tools.

A word of caution: be careful that the results are graphed in an appropriate manner. Most spreadsheet packages have very limited *x*-axis (horizontal) manipulation. Provided the *x*-axis data comes in regular steps, the results are acceptable. However, we have seen instances where very distorted graphs have been presented because of this problem.

Apart from the graphical interpretation of results, there are now several good packages that allow the analyst to enter the data graphically. It is a great advantage to be able to develop a one-line diagram of a network directly with the computer. All the relevant system components can be included. Parameter data still require entry in a more orthodox manner but by merely clicking on a component, a data form for that component can be made available. The chances of omitting a component are greatly reduced with this type of data entry. Further, the same system diagram can be used to show the results of some analyses.

An extension of the network diagram input is to make the diagram relate to the actual topography. In these cases, the actual routes of transmission lines are shown and can be superimposed on computer-generated geographical maps. The lines in these cases have their lengths automatically established and, if the line characteristics are known, the line parameters can be calculated.

These topographical diagrams are an invaluable aid for power reticulation problems, for example, the minimum route length of reticulation given all the points of supply and the route constraints. Other optimization algorithms include determination of line sizes and switching operations.

The analysis techniques can be either linear or non-linear. If successful, the non-linear algorithm is more accurate but these techniques suffer from larger data storage requirements, greater computational time, and possible divergence. There are various possible optimization techniques that can and have been applied to this problem. There is no definitive answer and each type of problem may require a different choice.

The capability chart represents a method of graphically displaying power system performance. These charts are drawn on the complex power plane and define the real and reactive power that may be supplied from a point in the system during steady-state operation. The power available is depicted as a region on the plane and the boundaries of the region represent the critical operating limits of the system. The best known example of a capability chart is the operating chart of a synchronous machine. The power available from the generator is restricted by limiting values of the rotor current, stator current, turbine power (if a generator), and synchronous stability limits. Capability charts have been produced for transmission lines and HV DC converters.

7.10   Protection

The need to analyze protection schemes has resulted in the development of protection co-ordination programs. Protection schemes can be divided into two major groupings: unit and non-unit schemes.

The first group contains schemes that protect a specific area of the system, i.e., a transformer, transmission line, generator or busbar. The most obvious example of unit protection schemes is based on Kerchief's current law – the sum of the currents entering an area of the system must be zero. Any deviation from this must indicate an abnormal current path. In these schemes, the effects of any disturbance or operating condition outside the area of interest are totally ignored and the protection must be designed to be stable above the maximum possible fault current that could flow through the protected area. Schemes can be made to extend across all sides of a transformer to account for the different currents at different voltage levels. Any analysis of these schemes is thus of more concern to the protection equipment manufacturers.

The non-unit schemes, while also intended to protect specific areas, have no fixed boundaries. As well as protecting their own designated areas, the protective zones can overlap into other areas. While this can be very beneficial for backup purposes, there can be a tendency for too great an area to be isolated if a fault is detected by different non-unit schemes. The most simple of these schemes measures current and incorporates an inverse time characteristic into the protection operation to allow protection nearer to the fault to operate first. While this is relatively straightforward for radial schemes, in networks, where the current paths can be quite different depending on operating and maintenance strategies, protection can be difficult to set and optimum settings are probably impossible to achieve. It is in these areas where protection software has become useful to manufacturers, consultants and utilities.

The very nature of protection schemes has changed from electromechanical devices, through electronic equivalents of the old devices, to highly sophisticated system analyzers (microprocessor-based relays or IEDs). They are computers in their own right and thus can be developed almost entirely by computer analysis techniques.

7.11   Other uses for load flow analysis

It has already been demonstrated that load flow analysis is necessary in determining the economic operation of the power system and it can also be used in the production of capability charts. Many other types of analyses require load flow to be embedded in the program, e.g.

- *Unbalanced three-phase analysis:*   As a follow-on from the basic load flow analysis, where significant unbalanced load or unbalanced transmission causes problems, a three-phase load flow may be required to study their effects. These programs require each phase to be represented separately and mutual coupling between phases to be taken into account. Transformer winding connections must be correctly represented and the mutual coupling between transmission lines on the same tower or on the same right-of-way must also be included.
- *Motor starting studies:*   Motor starting can be evaluated using a transient stability program but in many cases this level of analysis is unnecessary. The voltage dip associated with motor start-up can be determined very precisely by a conventional load flow program with a motor starting module.
- *Optimal power flow:*   Optimal power system operation requires the best use of resources subject to a number of constraints over any specified time period. The problem consists of minimizing a scalar objective function (normally a cost criterion) through the optimal control of a vector of control parameters. This is subject to the equality constraints of the load flow equations, inequality constraints on the control parameters, and inequality constraints of dependent variables and dependent functions. The programs to do this analysis are usually referred to as optimal power flow (OPF) programs.
- *Security assessment:*   Often optimal operation conflicts with the security requirements of the system. Load flow studies are used to assess security (security assessment). This can be viewed as two separate functions. First, there is a need to detect any operating limit violations through continuous monitoring of the branch flows and nodal voltages. Second, there is a need to determine the effects of branch outages (contingency analysis). To reduce this to a manageable level, the list of contingencies is reduced by judicial elimination of most of the cases that are not expected to cause violations.

From this the possible overloading of equipment can be forecast. The program should be designed to accommodate the condition where generation cannot meet the load because of network islanding.

The conflicting requirements of system optimization and security require that they be considered together. The more recent versions of OPF interface with contingency analysis and the computational requirements are enormous.

7.12   Extensions to transient stability analysis

Transient stability programs have been extended to include many other system components, including flexible AC transmission systems (FACTS) and DC converters.

Flexible AC transmission systems may be either shunt or series devices. Shunt devices usually attempt to control busbar voltage by varying their shunt susceptance. The device is therefore relatively simple to implement in a time domain program. Series devices may be associated with transformers. Stability improvement is achieved by injecting a quadrature component of voltage derived from the other two phases rather than by a tap changer, which injects a direct component of voltage. Fast acting power electronics can inject either direct voltage or a combination of both direct and quadrature voltage to help maintain voltage levels and improve stability margins.

DC converters for HV DC links and rectifier loads have received much attention. The converter controls are very fast acting and therefore a quasi steady-state (QSS) model can be considered accurate. That is, the model of the converter terminals contains no dynamic equations and in effect the link behaves as if it was in steady state for every time solution of the AC system. While this may be so some time after a fault has been removed, during and just after a fault the converters may well suffer from commutation failure or fire through. These events cannot be predicted or modeled with a QSS model. In this case, an appropriate method of analysis is to combine a state variable model of the converter, which can model the firing of the individual valves, with a conventional multi-machine transient stability program containing a QSS model. During the period of maximum disturbance, the two models can operate together. Information about the overall system response is passed to the state variable model at regular intervals. Similarly, the results from the detailed converter model are passed to the multi-machine model overriding its own QSS model. As the disturbance reduces, the results from the two different converter models converge and it is then only necessary to run the computationally inexpensive QSS model within the multi-machine transient stability program.

7.13 Voltage collapse

Steady-state analyses of the problem of voltage instability and voltage collapse are often based on load flow analysis programs. However, time solutions can provide further insight into the problem.

A transient stability program can be extended to include induction machines, which are associated with many of the voltage collapse problems. In these studies, it is the stability of the motors that are examined rather than the stability of the synchronous machines. The asynchronous nature of the induction machine means that rotor angle is not a concern, but instead the capability of the machines to recover after a fault has depressed the voltage and allowed the machines to slow down. The re-accelerating machines draw more reactive current, which can hold the terminal voltage down below that necessary to allow recovery. Similarly starting a machine will depress the voltage, which affects other induction machines, which further lowers the voltage.

However, voltage collapse can also be due to longer-term problems. Transient stability programs then need to take into account controls that are usually ignored. These include automatic transformer tap adjustment and generator excitation limiters, which control the long-term reactive power output to keep the field currents within their rated values.

7.14 SCADA

Supervisory control and data acquisition (SCADA) has been an integral part of system control for many years. A control center now has much real-time information available so that human and computer decisions about system operation can be made with a high degree of confidence.

In order to achieve high-quality input data, algorithms have been developed to estimate the state of a system based on the available online data (state estimation). These methods are based on weighted least squares techniques to find the best state vector to fit the scatter of data. This becomes a major problem when conflicting information is received. However, as more data becomes available, the reliability of the estimate can be improved.

7.15   **Power quality**

One form of poor power quality that has received a large amount of attention is the high level of harmonics that can exist and there are numerous harmonic analysis programs now available.

Recently, the harmonic levels of both currents and voltages have increased considerably due to the increasing use of non-linear loads such as arc furnaces, HV DC converters, FACTS equipment, DC motor drives, and variable AC motor speed control. Moreover, commercial sector loads now contain often unacceptable levels of harmonics due to widespread use of equipment with rectifier-fed power supplies with capacitor output smoothing (e.g. computer power supplies and fluorescent lighting). The need to conserve energy has resulted in energy-efficient designs that exacerbate the generation of harmonics. Although each source only contributes a very small level of harmonics, due to their small power ratings, widespread use of small non-linear devices may create harmonic problems that are more difficult to remedy than one large harmonic source.

Harmonic analysis algorithms vary greatly in their algorithms and features; however, almost all use the frequency domain. The most common technique is the direct method (also known as current injection method). Spectral analysis of the current waveform of the non-linear components is performed and entered into the program. The network data is used to assemble a system admittance matrix for each frequency of interest. This set of linear equations is solved for each frequency to determine the node voltages and, hence, current flow throughout the system. This method assumes the non-linear component is an ideal harmonic current source.

The next more advanced technique is to model the relationship between the harmonic currents injected by a component to its terminal voltage waveform. This then requires an iterative algorithm, which does require excursion into the time domain for modeling this interaction. When the fundamental (load flow) is also included, thus simulating the interaction between fundamental and harmonic frequencies, it is termed a harmonic power flow. The most advanced technique, which is still only a research tool, is the harmonic domain. In this iterative technique one Jacobian is built up that represents all harmonic frequencies. This allows coupling between harmonics, which occurs, for example, in salient synchronous machines, to be represented.

There are many other features that need to be considered, such as whether the algorithm uses symmetrical components or phase co-ordinates, or whether it is single or three phase. Data entry for single phase typically requires the electrical parameters, whereas three-phase analysis normally requires the physical geometry of the overhead transmission lines and cables and conductor details so that a transmission line parameter program or cable parameter program can calculate the line or cable electrical parameters.

The communication link between the monitoring point and the control center can now be very sophisticated and can utilize satellites. This technology has led to the development of systems to analyze the power quality of a system. Harmonic measurement and analysis has now reached a high level of maturity. Many different pieces of information can be monitored and the results over time stored in a database. Algorithms based on the fast Fourier transforms can then be used to convert this data from the time domain to the frequency domain. Computing techniques coupled with fast and often parallel computing allow this information to be displayed in real time. By utilizing the time-stamping capability of the global positioning system (GPS), information gathered at remote sites can be linked together. Using the GPS time stamp, samples taken exactly simultaneously can be fed to a harmonic state estimator which can even determine the position and magnitude of harmonics entering the system as well as the harmonic voltages and currents at points not monitored (provided enough initial monitoring points exist).

One of the most important features of harmonic analysis software is the ability to display the results graphically. The refined capabilities of present three-dimensional graphics packages have simplified the analysis considerably.

7.16   Finite element analysis

Finite element analysis is not normally used by power system engineers although it is a common tool of high voltage and electrical machine engineers. It is necessary, for example, where accurate machine representation is required. For example, in a unit-connected HV DC terminal the generators are closely coupled to the rectifier bridges. The AC system at the rectifier end is isolated from all but its generator. There is no need for costly filters to reduce harmonics. Models of the synchronous machine suitable for a transient stability study can be obtained from actual machine tests. For fast transient analysis, a three-phase generator model can be used but it will not account for harmonics.

A finite element model of the generator provides the means of allowing real-time effects such as harmonics and saturation to be directly included. Any geometric irregularities in the generator can be accounted for and the studies can be done at the design stage rather than having to rely on measurements or extrapolation from manufactured machines to obtain circuit parameters. There is no reliance on estimated machine parameters.

The disadvantages are the cost and time to run a simulation and it is not suitable at present to integrate with existing transient stability programs, as it requires a high degree of expertise. As the finite element model is in this case used in a time simulation, part of the air gap is left unmeshed in the model. At each time step the rotor is placed in the desired position and the missing elements in the air gap region formed using the nodes on each side of the gap.

7.17   Earthing

The safe earthing of power system equipment is very important, especially as the short-circuit capability of power systems continues to grow. Programs have been developed to evaluate and design earthing systems in areas containing major power equipment, such as substations and to evaluate the effects of fault current on remote, separately grounded equipment.

The connection to earth potential may consist of an earth mat of buried conductors, electrodes (earth rods), or both. The shape and dimensions of the electrodes, their locations, and the layout of an earth mat, plus the resistivity of the earth at different levels must be specified in order to evaluate the earth resistance. A grid of buried conductors and electrodes is usually considered to be all at the same potential. Where grid sections are joined by buried or aerial links, these links can have resistance allowing the grid sections to have different potentials. It is usual to consider a buried link as capable of radiating current into the soil.

Various methods of representing the fault current are available. The current can be fixed or it can be determined from the short-circuit MVA and the busbar voltage. A more complex fault path may need to be constructed for faults remote from the site being analyzed.

From the analysis, the surface potential over the affected area can be evaluated and, from that, step and touch potentials calculated. Three-dimensional graphics of the surface potentials are very useful in highlighting problem areas.

7.18   Other programs

There are too many other programs available to be discussed. For example, neither automatic generator control nor load forecasting has been included. However, an example of a small program that can stand alone or fit into other programs is given here.

In order to obtain the electrical parameters of overhead transmission lines and underground cables, utility programs have been developed. Transmission line parameter programs use the physical geometry of the conductors, the conductor type, and ground resistivity to calculate the electrical parameters of the line. Cable parameter programs use the physical dimensions of the cable, its construction, and its position in the ground. The results of these programs are usually fed directly to network analysis programs such as load flow or faults. The danger of errors introduced during transfer is thus minimized. This is particularly true for three-phase analyses due to the volume of data involved.

7.19   Further development of programs

Recently there has been a shift in emphasis in the types of program being constructed. Deregulation is making financial considerations a prime operating constraint. New programs are now being developed which assist in the buying and selling of energy through the electrical system.

Following on from the solution of the economic dispatch, 'time-of-use' pricing has been introduced into some power system operations. Under this system, the price of electricity at a given time reflects the marginal cost of generation at that time. As the marginal generator changes over time, so does the price of electricity.

The next stage is to price electricity not only on time but also on the place of use (nodal pricing). Thus, the cost of transportation of the energy from the producer to the user is included in the price. This can be a serious problem at present when power is exchanged between utilities. It will become increasingly common as the individual electricity producers and users set up contractual agreements for supply and use. A major problem at present is the lack of common agreement as to whether nodal pricing is the most appropriate mechanism for a deregulated wholesale electricity market. Clarification will occur as the structure of the industry changes.

Nodal pricing also takes into account other commercial and financial factors. These include the pricing of both generation and transmission constraints, the setting of a basis for transmission constraint hedges and for the economic dispatch of generation. The programs must be designed to give both the suppliers and consumers of energy the full opportunity costs of the operation of the power system.

> *Inherent in nodal pricing must be such factors as marginal cost pricing, short run price, and whether the price is* ex ante *(before) or* ex post *(after) the event. Thus far, the programming effort has concentrated on real power pricing but the cost of reactive power should also eventually be included.*

The changes in the operation of power systems, which are occurring throughout the world at present, will inevitably force changes to many of the programs in use today and, as shown above, new programs will emerge.

7.20   Program suites

As more users become involved with a program, its quirks become less acceptable and it must become easy to use, i.e., user friendly. Second, with the availability of many different types of program, it became important to be able to transfer the results of one program to the input of another. If the user has access to the source code, this can often be done relatively quickly by generating an output file in a suitable format for the input of the second program. There has, therefore, been a great deal of attention devoted to creating common formats for data transfer as well as producing programs with easy data entry formats and good result processing capabilities.

Many good 'front end' programs are now available which allow the user to quickly write an analysis program and utilize the built-in IO features of the package. There are also several good general mathematical packages available. Much research work can now be done using tools such as these. The researcher is freed from the chore of developing algorithms and IO routines by using these standard packages. Not only that, extra software is being developed which can turn these general packages into specialist packages. It may well be that before long all software will be made to run on sophisticated developments of these types of package and the stand-alone program will fall into oblivion.

7.21 Conclusions

There are many more programs available than can be discussed here. Those that have been discussed are not necessarily more significant than those omitted. There are programs to help you with almost every power system problem you have and new software is constantly becoming available to solve the latest problems.

Make sure that programs you use are designed to do the job you require. Some programs make assumptions that give satisfactory results in most cases but may not be adequate for your particular case. No matter how sophisticated and friendly the program may appear, it is the algorithm and processing of data that are the most important parts. As programs become more complex and integrated, new errors (regressions) can be introduced. Wherever possible check the answers and always make sure they feel right.

Last, but not least, remember the golden rule of all computer programs and simulations: GIGO – Garbage In, Garbage Out. Make absolutely sure that all information is input into the program correctly, and that it is defined in the right context and units. For example, a cable length defined as 300 miles instead of 300 m will make quite a substantial difference to the result of analyses.

8

New era of power system automation

8.1   Definition of the term

Power system automation can be defined as a system for managing, controlling and protecting an electrical power system. This is accomplished by obtaining real-time information from the system, having powerful local and remote control applications and advanced electrical protection. The core ingredients of a power system automation system are local intelligence, data communications and supervisory control and monitoring.

Note: Power system automation is also referred to as substation automation.

8.2   What is power system automation?

Power system automation may be best described by referring to Figure 8.1.

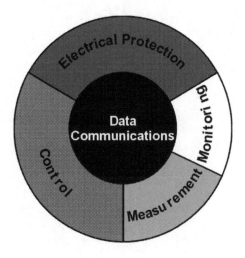

Figure 8.1
Functional structure of power system automation

Power system automation, by definition, consists of the following main components:

- Electrical protection
- Control
- Measurement
- Monitoring
- Data communications.

Electrical protection

Electrical protection is still one of the most important components of any electrical switchgear panel, in order to protect the equipment and personnel, and to limit damage in case of an electrical fault.

Electrical protection is a local function, and should be able to function independently of the power system automation system if necessary, although it is an integral part of power system automation under normal conditions. The functions of electrical protection should never be compromised or restricted in any power system automation system.

Control

Control includes local and remote control. Local control consists of actions the control device can logically take by itself, for example bay interlocking, switching sequences and synchronizing check. Human intervention is limited and the risk of human error is greatly reduced.

Local control should also continue to function even without the support of the rest of the power system automation system.

Remote control functions to control substations remotely from the SCADA master(s). Commands can be given directly to the remote-controlled devices, for example open or close a circuit breaker. Relay settings can be changed via the system, and requests for certain information can be initiated form the SCADA station(s). This eliminates the need for personnel to go to the substation to perform switching operations, and switching actions can be performed much faster, which is a tremendous advantage in emergency situations.

A safer working environment is created for personnel, and huge production losses may be prevented. In addition, the operator or engineer at the SCADA terminal has a holistic overview of what is happening in the power network throughout the plant or factory, improving the quality of decision-making.

Measurement

A wealth of real-time information about a substation or switchgear panel is collected, which is typically displayed in a central control room and/or stored in a central database.
Measurement consists of:

- Electrical measurements (including metering) – voltages, currents, power, power factor, harmonics, etc.
- Other analog measurements, e.g. transformer and motor temperatures
- Disturbance recordings for fault analyses.

This makes it unnecessary for personnel to go to a substation to collect information, again creating a safer work environment and cutting down on personnel workloads. The huge amount of real-time information collected can assist tremendously in doing network studies like load flow analyses, planning ahead and preventing major disturbances in the power network, causing huge production losses.

Note: The term 'measurement' is normally used in the electrical environment to refer to voltage, current and frequency, while 'metering' is used to refer to power, reactive power, and energy (kWh). The different terms originated due to the fact that very different instruments were historically used for measurement and metering. Nowadays the two functions are integrated in modern devices, with no real distinction between them; hence the terms 'measurement' and 'metering' are used interchangeably in the text. Accurate metering for billing purposes is still performed by dedicated instruments.

Monitoring

- Sequence-of-event recordings
- Status and condition monitoring, including maintenance information, relay settings, etc.

This information can assist in fault analyses, determining what happened when, where and in what sequence. This can be used effectively to improve the efficiency of the power system and the protection. Preventative maintenance procedures can be utilized by the condition monitoring information obtained.

Data communication

Data communication forms the core of any power system automation system, and is virtually the glue that holds the system together. Without communications, the functions of the electrical protection and local control will continue, and the local device may store some data, but there can be no complete power system automation system functioning. The form of communications will depend on the architecture used, and the architecture may, in turn, depend on the form of communication chosen.

8.3   Power system automation architecture

Different architectures exist today to implement the components of power system automation in practice. It is important to realize that not one single layout can exclusively illustrate a power system automation system. However, the most advanced systems today are developing more and more toward a common basic architecture. This architecture is illustrated in Figure 8.2.

The modern system consists of three main divisions.

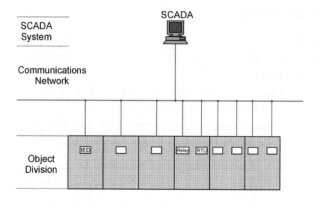

Figure 8.2
Basic architecture of power system automation

Object division

The object division of the power system automation system consists of intelligent electronic devices (IEDs), modern, third generation microprocessor-based relays and/or remote terminal units (RTUs). The PLCs continue to play an important role in some systems. They receive analog inputs from the current transformers (CTs), voltage transformers (VTs) and transducers in the various switchgear panels, as well as digital inputs from auxiliary contacts, other field devices or IEDs, or the SCADA master. They are able to perform complex logical and mathematical calculations and provide an output either to the SCADA master, other field instruments or IEDs, or back to the switchgear to perform some command, for example open a circuit breaker.

The object division consists of the process level (field information from CTs, VTs, etc.) and the bay level (local intelligence in the form of IEDs, RTUs, etc.).

The communications network

The communications network (comms network for short) is virtually the nervous system of power system automation. The comms network ensures that raw data, processed information and commands are relayed quickly, effectively and error-free among the various field instruments, IEDs and the SCADA system. The physical medium will predominantly be fiber-optic cables in modern networks, although some copper wiring will still exist between the various devices inside a substation.

The comms network need to be an 'intelligent' subsystem in its own right to perform the functions required of it, and is not merely a network of fiber-optic and copper wiring.

The communication network serves as the interface between the bay level and the SCADA station level, which might be a SCADA master station in the substation itself, or remotely in a central control room.

SCADA master

The SCADA master station(s) forms the virtual brain of the power system automation system. The SCADA master receives data and information from the field, decides what to do with it, stores it (directly or after some form of processing), and issues requests and/or commands to the remote devices. Therefore, the SCADA master is effectively in control of the complete power system automation system.

Nowadays, a SCADA master consists simply of an advanced, reliable PC or workstation (with its peripheral and support hardware) and a SCADA software package. (In contrast with a few years ago when SCADA systems used to run on big mainframe computers or some form of complex proprietary hardware.)

A SCADA master station may be installed in each substation of a power transmission network (station level), with all the substation SCADA stations forming part of a LAN or WAN (network level); or one SCADA master station may be directly in control of several substations, eliminating the station level.

Appendix

Name	P1 (kW)	Q1 (KVA)	Pf cos(theta)	S1 (kVA)	Voltage (kV)	I1 (A)	In Breaker (A)	I Fault Breaker (kA)	Cable Size Volt Drop (mm²)	Cable Size Load Current (mm²)	Cable Size Fault Level (mm²)	Actual Cable Size (mm²)
			sub #4									
P201	375.0		0.80	468.8	6.60							
P202	550.0		0.85	647.1	6.60							
P203	200.0		0.80	250.0	6.60							
tfr C			0.60	1000.0	6.60							
tfr D			0.60	1000.0	6.60							
tfr E			0.60	1000.0	6.60							
P804	320.0		0.80	400.0	6.60							
P805	1500.0		0.85	1764.7	6.60							
P806	250.0		0.80	312.5	6.60							
Total / L4 / 0 / R4 / busbars				5843.0	6.60							
			sub #3									
Cap	0.0	−2400.0	0.00		6.60							
P301	480.0		0.80		6.60							
P302	900.0		0.85		6.60							
P303	300.0		0.80		6.60							
tfr F			0.60	1250.0	6.60							
tfr I			0.60	1250.0	6.60							
P306	375.0		0.80		6.60							
P307	461.0		0.80		6.60							
P308	2000.0		0.85		6.60							
Cap	0.0	−2400.0	0.00		6.60							
L2 sub 4 feeder					6.60							
Total / L8 / 0 / R8 / busbars					6.60							

System Fault Levels

	At 132 kV	At 33 kV with 1 tfr	At 33 kV with 2 tfrs	At 33 kV without source impedance – just 2 tfrs	At 6.6 kV with 1 tfr	At 6.6 kV with 2 tfrs	At 6.6 kV without source impedance – just 2 tfrs
Base power in MVA	100.0		100.0			100.0	
Voltage level in kV	132.0		33.0			6.6	
Base impedance in ohms							
Base current in A							
Source impedance in pu							
OH Line 10 km wolf							
40 MVA tfr							
Cable (185 mm^2 × 4–5 km)							
10 MVA tfr							
Total impedance							
Fault level in p.u.							
Fault level in A at 132 kV							
Fault level in A at 33 kV							
Fault level in A at 6.6 kV							

Name	P$_{nom}$ (kW) Nominal	Q$_{nom}$ (kVAr) Nominal	Pf cos(theta)	Theta	S$_{nom}$ (kVA)	Loading (Pu)	P$_{actual}$ (kW)	Q$_{actual}$ (kVAr)	S$_{actual}$ (kVA)	Voltage (kV)	I$_{actual}$ (A)
						sub #4					
P201	375.0		0.80			0.80				6.60	
P202	550.0		0.85			0.80				6.60	
P203	200.0		0.80			0.80				6.60	
tfr C			0.60		1000.0	0.45				6.60	
tfr D			0.60		1000.0	0.45				6.60	
tfr E			0.60		1000.0	0.45				6.60	
P804	320.0		0.80			0.80				6.60	
P805	1500.0		0.85			0.80				6.60	
P806	250.0		0.80			0.80				6.60	
Total										6.60	
						sub #3					
cap	0.0	−2400.0	0.00		0.0	1.00				6.60	
P301	480.0		0.80			0.80				6.60	
P302	900.0		0.85			0.80				6.60	
P303	300.0		0.80			0.80				6.60	
tfr F			0.60		1250.0	0.45				6.60	
tfr I			0.60		1250.0	0.45				6.60	
P306	375.0		0.80			0.80				6.60	
P307	461.0		0.80			0.80				6.60	
P308	2000.0		0.85			0.80				6.60	
cap	0.0	−2400.0	0.00		0.0	1.00				6.60	
L2 sub 4 feeder										6.60	
Total										6.60	
Total (uncompensated)										6.60	

Index